Observing Our Universe

David Michalets

Self-published on **July 30, 2020**

Table of Contents

Introduction 4

1 Red Shifts 8

2 Gravitational Waves . . . 76

3 Relativity 113

4 Cosmology Concerns . 159

5 Cosmology in Chaos . 204

6 Final Conclusion 218

7 References 222

Introduction

This book is about observing our universe, from on or near our Earth.

When we look up at the night sky, we can forget we are observing from only one view. If we had another view for comparison, like from another galaxy, our understanding of our universe would be different.

We are observing our universe with only one context, near our Earth. We are observing with this particular line of sight, when measuring a spectrum.
The resulting misperception is everything is moving away from us. This illusion can be explained.

The Doppler Effect gets a thorough explanation for galaxies, quasars, and stars.

We can observe a gravitational wave indirectly, only by its proposed effect on the Earth's surface.

Some consider the detection of gravitational waves to be a confirmation of Einstein's relativity.

We are unable to observe a black hole, or dark matter and dark energy.

We could not observe the big bang which was in the past.

Einstein's Theory of Special Relativity defined a special observer, one who is moving through a gravitational field. We are not special because we are not moving past the distant objects while here on Earth).

The context for every observation is important.

The foundations of both astrophysics and cosmology rely on a spectrum analysis and on Einstein's theory of relativity.

Every claim made in astrophysics or cosmology requires evidence to be acceptable.

This is a brief summary of the 7 sections:

1) A spectrum analysis can try to measure relative motion of a celestial object. Our line of sight from Earth affects that analysis. A red shift is an important result.

The illusion is everything is moving away, by a red shift.

2) Gravitational waves are detected indirectly. It is crucial to consider the Earth for any wave detection before assigning the wave to an astrophysical source.

3) There are other concerns with the cosmological application of relativity beyond gravitational waves. Relativity is based on the moving observer's reference frame called space-time. That special observer's context is crucial.

Relativity was apparently confirmed as a valid replacement for Newton's force of gravity. That claim of confirmation must be verified.

4) There are other concerns with cosmology beyond the above sections.

These concerns include dark matter which remains elusive and the big bang with its CMB in debate.

Cosmology is known to be in a crisis because attempts to agree on a value of Hubble's constant using several methods have failed to converge on a single value. This constant is critical to the assumed universe expansion and its assumed acceleration. The failure to agree has been called a crisis by the cosmologists involved in this pursuit.

5) Based on the red shift illusion, the known crisis, and the conclusions in the above sections, cosmology is in chaos.

6) Final conclusion results from the summation of the above sections.

The conclusions in these sections affect the foundation of astrophysics and cosmology.

An observer must properly account for their method of observation to obtain a correct conclusion.

7) All external references in the book have links available as directed here.

1 Red Shifts

1.1 Introduction to Red Shifts

A red shift in a spectrum is assumed to reflect a velocity. However, sometimes a red shift is not a velocity. A detailed description of the Doppler Effect for both emission and absorption is provided. The correct and wrong methods of a spectrum analysis are also described.

Sometimes the velocity is used to calculate a distance to the light source. Distance measurements are also described.

There are two fundamental assumptions for how objects can exhibit the Doppler Effect.

The fundamental equations for the Doppler Effect are explained.

1.2 Terminology

a) Light and wavelengths

A spectrum is the entire range of wavelengths in electromagnetic radiation where light is the visible range. The ultraviolet and infrared ranges are not visible to the human eye but they are in the Sun's radiation. Because this topic is about the visible stars and galaxies, the word light is often used for the entire spectrum, including those frequency ranges not visible.

Electromagnetic radiation is the propagation of synchronized, perpendicular electric and magnetic fields. The propagation has a defined rate of oscillation measured as either a frequency or a wavelength.
The wavelength is usually measured in either nanometers (10^{-9} m) or Angstroms (10^{-10} m or 0.1 nm). The velocity of this propagation has been measured in a vacuum using our standard definition for time and this measured value is called the constant c. This measurement also defined the standard unit of 1 meter. The velocity of propagation is reduced in a medium, defined by the medium's diffraction index.
Light transmits energy proportional to its frequency so the constant c appears in some physics equations involving energy.

Quantum physics defined a theoretical particle called a photon to refer to a single wavelength.

In this book, wavelength is used because a spectrum analysis uses specific numerical values. Using photon instead of wavelength only introduces possible confusion. Photon will not be used in this topic's original content. When photon is in a reference excerpt, wavelength can be substituted for photon for consistency.

b) Fraunhofer Lines

This description provides background for many terms and their use in a spectrum analysis.

Excerpt from Wikipedia:

In 1814, Fraunhofer independently rediscovered the [dark] lines and began to systematically study and measure the wavelengths where these features are observed. He mapped over 570 lines.

About 45 years later Kirchhoff and Bunsen noticed that several Fraunhofer lines coincide with characteristic emission lines identified in the spectra of heated elements. It was correctly deduced that dark lines in the solar spectrum are caused by absorption by chemical elements in the solar atmosphere. Some of the observed features were identified as telluric lines originating from absorption by oxygen molecules in the Earth's atmosphere.

Because of their well–defined wavelengths, Fraunhofer lines are often used to characterize the refractive index and dispersion properties of optical materials.

(Excerpt end)

c) atom's characteristic wavelengths

Calcium and hydrogen are the most frequently observed atoms in the spectrum of a distant galaxy or quasar.

The calcium atom is important because a galaxy can have its ion's pair of calcium absorption lines at 3934 and 3969 Angstroms in its spectrum when a calcium ion is in the line of sight to the galaxy. A red or blue shift of this pair of lines indicates the relative velocity of the ion. The neutral calcium atom has a different pair of wavelengths. Nearly all matter in the universe is plasma, or it has an electrical charge. That includes electrons (-), protons(+), and ions (+) which are atoms having lost one or more electrons.

Hydrogen is the most common element in the universe; it is also the simplest having only one proton and one electron.

Excerpt from Wikipedia:

In physics, the Lyman-alpha line is a spectral line of hydrogen, or more generally of one-electron ions, in the Lyman series, emitted when the electron falls from the n = 2 orbital to the n = 1 orbital, where n is the principal quantum number. In hydrogen, its wavelength of 1215.67 angstroms corresponding to frequency of 10^{15} hertz, places the Lyman-alpha line in the ultraviolet part of the electromagnetic spectrum, which is absorbed by air. Lyman-alpha astronomy must therefore ordinarily be carried out by satellite-borne instruments, except for extremely distant sources whose red shifts allow the hydrogen line to penetrate the atmosphere.

(Excerpt end)

This wavelength is important because a quasar can have this emission line in its spectrum. A shift of this emission line wavelength indicates the relative velocity of the atom.

d) Doppler Effect

Excerpt from Britannica:

Doppler effect, the apparent difference between the frequency at which sound or light waves leave a source and that at which they reach an observer, caused by relative motion of the observer and the wave source. This phenomenon is used in astronomical measurements.

(Excerpt end)

The Doppler Effect is observed by the entire spectrum of the light source being shifted in proportion to the source's velocity in that direction.

The Doppler Effect occurs only at the moment of radiation emission or at the moment of radiation absorption when the motion of the object at that instant affects the spectrum.

e) Synchrotron Radiation

Excerpt from Wikipedia:

Synchrotron radiation, electromagnetic energy emitted by charged particles (e.g., electrons and ions) that are moving at speeds close to that of light when their paths are altered, as by a magnetic field. It is so called because particles moving at such speeds in a variety of particle accelerator that is known as a synchrotron produce electromagnetic radiation of this sort.

Many kinds of astronomical objects have been found to emit synchrotron radiation as well. High-energy electrons spiraling through the lines of force of the magnetic field around the planet Jupiter, for example, give off synchrotron radiation at radio wavelengths. Synchrotron radiation at such wavelengths and at those of visible and ultraviolet light is generated by electrons moving in the magnetic field associated with the supernova remnant known as the Crab Nebula. Radio emissions of the synchrotron variety also have been detected from other supernova remnants in the Milky Way Galaxy and from extragalactic objects called quasars.

(Excerpt end)

f) Thermal Radiation

Excerpt from Wikipedia:

Thermal radiation is electromagnetic radiation generated by the thermal motion of particles in matter. All matter with a temperature greater than absolute zero emits thermal radiation.

If a radiation object meets the physical characteristics of a black body in thermodynamic equilibrium, the radiation is called blackbody radiation. Planck's law describes the spectrum of blackbody radiation, which depends solely on the object's temperature. Wien's displacement law determines the most likely frequency of the emitted radiation, and the Stefan–Boltzmann law gives the radiant intensity.

(Excerpt end)

Thermal radiation is also one of the fundamental mechanisms of heat transfer. Conduction between adjacent solid objects is another.

Its spectrum is characterized by a frequency distribution, with the frequency having the highest intensity related to the object's temperature.

The frequency distribution affects whether it is visible. A cool temperature won't be. When warmer the increasing infrared intensity can be felt as heat or warmth but not seen. A rising temperature will become visible as red. When even hotter the mix of color frequencies can result in "white hot." Our Sun is hot enough to generate the ultraviolet frequency which is not visible but can affect the eyes and skin.

Our white Sun can appear yellow when overhead due to the frequency distribution after the light passes through our atmosphere. The atmosphere can also cause a color change between sun rise and sun set, and it causes the sky to be blue.

Important note about frequencies:

Thermal radiation typically spans from infrared to ultraviolet frequencies. Our Sun's thermal radiation, seen as light, is in this frequency range

Most emission lines from atoms range from visible to ultraviolet frequencies. As a general rule, any frequencies measured outside of this range, like radio at the low end, and X-ray or gamma ray at the high end, were emitted by a source of synchrotron radiation.

A black hole violates this general rule because the hot accretion disk is claimed to emit X-rays but that requires an impossible temperature.

g) AGN

The Active Galactic Nucleus, AGN, is the source of intense electromagnetic radiation from the core of a galaxy or quasar. This common term is used, but no theories of the AGN mechanism are provided. They are not necessary for the details in this book.

The emphasis is on how we interpret observations from Earth.

h) IGM

The Inter-galactic Medium, IGM, is the space between galaxies. This space is not a perfect vacuum but has particles at a low density. Some of these particles are critical to the analysis of light passing through the IGM. A consistent behavior within the IGM has been found through many observations.

i) Star Type

Astronomers have defined a number of types for stars roughly defined by their measured surface temperature. Those details of classification are not crucial here.
Both a star's spectrum and its brightness are crucial to understand that star.

However it is extremely rare to find a star's spectrum public. This makes it impossible for others to learn from the measurement. If the initial analysis leading to its classification had a mistake, it is impossible to correct it when having no data

j) Cepheid

A Cepheid is a variable star whose brightness changes in a consistent manner. This observation is called a luminosity curve and can be described simply as having a peak brightness, or luminosity, followed by a dimming over a number of days, then the star brightens to the peak again over a few days and the cycle repeats. This repeating pattern is called a luminosity curve.
There are only a few star types which can exhibit this behavior. Cepheids are among the hottest star types. The observed repeatable luminosity curves have variations so there are several types of variable stars.

The astronomer must match an observed variable star to a known combination of star type and its luminosity curve.
Dust in the line of sight can cause dimming of the star's light. Cepheids have a margin of error based on these factors.
However, they remain the dominant star type which offers a predictable pattern to serve this function.

Section 1.10 below describes a behavior of a variable star like a Cepheid. The common explanation for a pulsating variable star is called the kappa mechanism. It claims the entire star expands and contracts during the luminosity cycle. The spectrogram in section 1.10 reveals this is a temperature cycle.

A star is visible by its thermal radiation. If its surface is changing brightness then it must be changing its temperature. Whether the star's volume changes during the cycle is irrelevant. Apparently the temperature changes had been wrongly taken to be motions of the surface, because the kappa mechanism description implies a changing motion of the surface.

A stellar spectrum being public is extremely rare. That figure from a public paper is the only spectrum this author could find for a variable star.

A supernova is an extreme brightening of a star before it dims away.

Attempts to use a supernova's peak brightness as a similar metric have failed because each is unique with no distinct repeatable pattern. A reliable metric must be repeatable.

There are different cycles among several star types so there are several types of variable stars. When these stars were found in our Milky Way, their distances could be calculated by the parallax method.
Their consistency enables their use as a distance measurement tool.
Despite their margin of error, Cepheids provide a practical distance measurement to the galaxy where they reside. Astronomers must accept some uncertainty using this method. Sometimes, galaxy distances have a stated margin of error. Whether that stated margin comes from a Cepheid is never identified.

k) Galaxy

A galaxy can be simply described as a very large group of stars and associated matter.

Sometimes the galaxy has arms which spiral out from the center so they are called spiral galaxies.
Other galaxies don't rotate so they have a spherical shape which can be somewhat flatter in the shape of an ellipse, so the variety is called elliptical galaxies.
A lenticular galaxy is one in between; It has the central bulge of stars but shows only a edge-on disk of dust which does not rotate.
There are some galaxies which have no definite shape so they are called irregular galaxies.
Most know our Milky Way is a spiral galaxy. Because our Sun is in one of spiral arms, telescopes were used to find its structure.
The Magellanic Clouds visible from the Southern Hemisphere once considered irregular galaxies but further analysis found some rotation so both are a special spiral galaxy. Those two are our closest galaxies.
There rather small galaxies, so they are dwarf galaxies. Some may have a shape so there is a dwarf elliptical galaxy type.

The Milky Way has several small galaxies which appear to be tied by gravity to the Milky Way so they are called satellite galaxies.

There are other large groups of stars called a globular cluster. The name comes from their spherical shape. Globular clusters are similar to small elliptical galaxies because they don't rotate. Sometimes there is ambiguity like between a nearby dwarf elliptical galaxy or a globular cluster.

Globular clusters are always associated with a much larger galaxy, which can be any of the galaxy types mentioned above.
Whenever galaxies form large clusters there is nearly always at least one giant elliptical galaxy accompanied by any number of the other galaxy types.

l) Quasar

A quasar is short for quasi-stellar object. The quasar has an AGN surrounded by clouds which dim the visible light but can remain bright in radio and X-ray. The quasar always has a red shifted hydrogen emission line. An unusual object called a BL Lac object is like the quasar but lacks the hydrogen emission line. Both have an AGN generating synchrotron radiation. Their common AGN was the conclusion in a study of both by BeppoSAX in 2008.

This study is found with:
"BeppoSAX OBSERVATIONS OF SYNCHROTRON X-RAY EMISSION FROM RADIO QUASARS"

Quasars can have 2 red shift values so every time its spectrum is analyzed that result must be recognized.

The quasar spectrum is characterized by many emission lines. Nearly all of them are from the ions moving toward the AGN. When each captures an electron, the electron must fall into an appropriate orbit, or energy state, for that element. That change in the internal energy state is accompanied by the radiation of that energy as an emission line. The ion is moving so the emission line is red shifted by its velocity.

An analysis of the various emission lines can find magnesium, oxygen, sodium, and others.

These ions are moving at the same velocity in the electric field of this AGN,

This velocity of the ions is the lower of the 2 red shifts.

A quasar also exhibits another, higher red shift value. The analysis should find it. There are 2 ways to miss it:

1) it is not apparent among the many other lines, or

2) it is in far infrared by a very high red shift and the spectrograph does not show the very long wave lengths.

A high velocity proton can also move toward this AGN but at a velocity unrelated to the cloud of ions around the AGN.
When this proton captures an electron from this AGN, like the metallic ions are doing, the electron must fall into the correct orbit or energy state. That change in internal energy state is accompanied by the radiation of that energy as an emission line.

Because this proton is moving toward the AGN, or away from us (the observer), this emission line is red shifted by the proton's velocity.

Quasars have been measured with a red shift > 7. That is never the quasar's velocity.

Relativity incorrectly claims matter cannot move at a velocity greater than light. That was an assumption with no evidence to justify it. A quasar is the evidence proving relativity is wrong with this assumption.

If the person doing the analysis of the quasar spectrum does not find the hydrogen emission line among the other emission lines, the result is a lower red shift than another person might measure using that line.

However, none of red shift values matter for a quasar. All are from either protons or ions. The red shift never applies to the actual quasar.

The author's research into quasars (including the analysis of a quasar spectrum) is available on his web site, cosmologyview. That explanation of the origin of an ignored value is not important for this book.

m) Star or Sun

A star has been described as a sphere of gas or plasma, held together by gravity and powered by fusion. This is sometimes called the Gaseous Sun.

The internal processes in a star are not important to this book's conclusions. However, a number of terms for a star's features are used so their reference is needed. One term is condensed matter.

There is an alternate theory of a star's internal structure called the Liquid Metallic Hydrogen Solar Model, or LMHSM.

The important difference is the Sun is composed of condensed matter, not a gas.

This theory became public with the Dr. Pierre-Marie Robitaille paper in 2013 titled:

"Forty Lines of Evidence for Condensed Matter - The Sun on Trial: Liquid Metallic Hydrogen as a Solar Building Block"

Dr. Robitaille has a YouTube channel, Sky Scholar, with many videos about stars and this model.

This model matches the helio-seismology data and all the solar observations, like limb darkening. The gaseous sun model lacks sufficient evidence, when unable to explain all observations, including its liquid surface, different rotation rates by latitude, its hot corona, solar wind acceleration, and many more.

From Wikipedia: here is the Sun's thermal radiation:

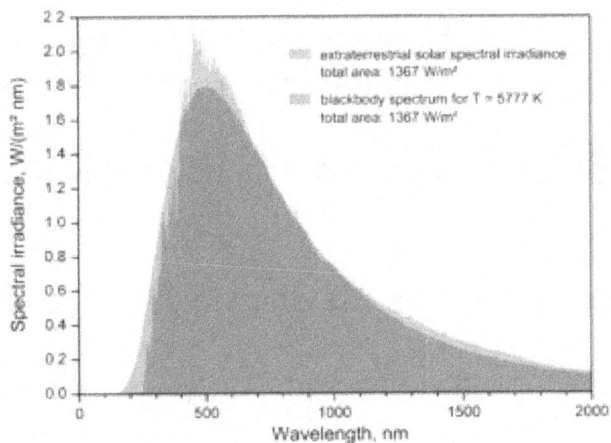

The caption:

The effective temperature, or black body temperature, of the Sun (5777 K) is the temperature a black body of the same size must have to yield the same total emissive power.

The external behaviors of the Sun are described by a new theory called the Electric Sun.

The solar surface has been duplicated in a laboratory by the SAFIRE project, effectively demonstrating this model.

There are several YouTube videos describing this theory and this experiment, but here are two:

"Donald Scott The Electronic Sun Electric Universe 2012"

"Donald Scott: SAFIRE and the Electric Sun | Space News"

Astrophysics is beginning a transition to a different explanation for a star. That is not crucial to the conclusion in this book.

This transition is mentioned here because evidence should be required in physics. These new solar theories have the evidence to verify them.

1.3 Fundamental Assumptions

There are two fundamental assumptions for the Doppler Effect.

First,

Absorption and emission lines come from atoms in the line of sight.

Observations from Earth's surface must account for the known atoms in the atmosphere.

   For a star within the Milky Way these atoms are usually near the star.
In many cases, these atoms are moving with the star behind them, just like the many absorption lines observed in the solar spectrum. When independent, the atom's apparent motion cannot be applied to the light source.
The Solar spectrum has emission lines from ions in the solar corona, well above the solar surface.

Second,

This Doppler Effect applies only to a single object because the Doppler Effect involves the kinetic energy of the light source so the effect can apply only to an individual mass and its velocity. This Doppler Effect shift occurs only at the moment of either transmission or absorption.

1.4 Fundamental Calculations

The kinetic energy of the object is involved in the energy transfer at the moment of the Doppler Effect.

This is the simple calculation of z.

First the velocity, called v here, of the source is compared to the velocity of light by dividing that value by the velocity of light, called the constant c.

The result is called z by convention.

The simple equation is $z=v/c$, making sure the units are the same (usually km/s).

The spectrum being emitted by the radiation source in the direction of travel has its wavelengths reduced or toward the blue end of the spectrum; this is called a blue shift.

The spectrum being emitted by the radiation source in the direction opposite of travel has wavelengths increased or toward the red end of the spectrum; this is called a red shift.

The shift in a spectrum due to the motion of the light source is a simple equation,
where EWL is the emission wavelength,

NWL is the new wavelength, so:
NWL = EWL + (EWL multiplied by z)

where the z is the factor for the change in the new wavelength from that originally emitted; z is positive for a red shift or negative for a blue shift.

It is very important to note:

The Doppler Effect changes the spectrum only at the single object (defined above), based on the object's velocity and direction.

The Doppler Effect does not change the total energy emitted by the light source. The single light source is generating a sphere of radiated energy in a continuum. In the direction of travel the energy is slightly increased by the wavelength reduction. In the direction opposite of travel the energy is slightly decreased by the wavelength increase. The energy change in one direction is exactly matched by the change in the other direction. The sphere of radiated energy is no longer uniform due to the motion of the light source but the total energy remains the same.

Depending on the line of sight the object could have either a blue or red shift.

The amount of wave length shift is driven by both the velocity and the angle between the light and the observer.
The Doppler Effect does not violate the conservation of energy, a basic principle of thermodynamics, because there is no energy gained or lost during the moment.

1.5 Doppler Effect around a sphere of Radiation

The sight source is radiating a fixed amount of energy per wavelength. A maintained source repeats each wavelength being emitted, though in reality the radiated energy is usually a continuum of wave lengths, like from a star and its photosphere. An atom can emit one or more wave lengths depending on the element's electron configuration.

In the case of an atom changing to a lower energy state, a characteristic wave length EWL is emitted with that energy. This is a transfer from internal energy to radiated energy.

The previous section 1.4 Fundamental Equations had acronyms and definitions for Emitted Wave Length, and New Wave Length which is the result of relative velocity measured as z, using both its sign and value.

The calculations are simple and found in many references.

The 2 scenarios are simple:

1) the light source is not moving,

2) the light source is moving.

The frequency for a wave length is simply the constant c divided by the wave length.

The energy being radiated is simply Planck's constant multiplied by the frequency.

This simple image is a circle with only 6 rays but the radiation is really a spherical continuum of energy from the source at the center.

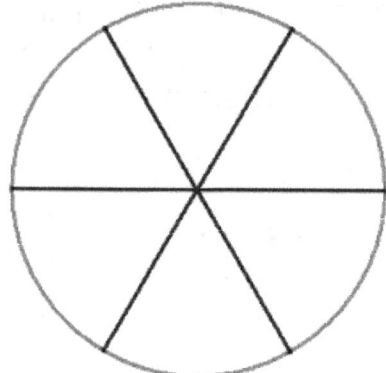

Treating the simple image like a clock, there are rays at 1, 3, 5, 7, 9, and 11 o'clock.

With scenario (1) or not moving: this energy is uniform in all directions, so the same wavelength is emitted in all directions. This is a sphere of energy propagating from the source.

The propagation continues until it is absorbed by an object.

With scenario (2) or moving: this same energy is being emitted from the source but the wavelength is changing around the spherical continuum from the source.

NWL is the New emitted Wave Length in a particular direction based on z at that point at the source in the continuum.

If the source is moving in direction 3 o'clock or 3 o, then the NWL will be decreased (by negative z) in this direction because this ray is in the direction of travel. Wave lengths are shorter.
If the source is moving in direction 9 o, then the NWL will be increased (by positive z) because this ray is in the opposite direction of travel. Wave lengths are longer.
The change values will be the same at 3 o and 9 o but with opposite sign.
Both the z value and polarity vary around the spherical continuum.
The change values at 5 o and 11 o are opposites, and at 1o and 7 o.

The energy being radiated from the light source is maintained regardless of direction of motion but the distribution of energy around the sphere of radiation has a smooth wave length change gradient in its distribution around the sphere.

The image is a 2-D circle. In reality, the radiation is an expanding 3-D sphere with the (moving) object at its center.

An atom in space absorbing light from a distant source will perform the same spherical pattern of absorbing a wavelength based on its velocity or kinetic energy and its direction relative to the incoming light path, so in a similar manner the z varies in value and sign within the spherical continuum around this atom.

If the atom is moving in direction 3 o relative to the light path, then the NWL will be increased (by positive z) because this absorbed ray is in the opposite direction of travel. Wave lengths are longer, or red shifted.
The change values will be the same at 3 o and 9 o but with an opposite sign.

The change values at 5 o and 11 o are opposites, and also at 1o and 7 o.

The Doppler Effect and an emission line are **always** a conservation of energy.

A star with its sphere of radiation will have the wavelengths uniformly changed around the spherical continuum, based on the star's motion, but the total energy being emitted is maintained, or conserved.

The atom will absorb a particular wavelength based on both its velocity and the light path direction relative to the atom's direction.

An absorption line is a transfer of energy. The energy of the original wave length matches the sum of the energy transferred to the atom's internal state change plus the energy of the wave length absorbed by the atom.

The Doppler Effect and an absorption line are **always** a conservation of energy

In practice, given the scale of the universe and we are observing the distant object in our line of sight, typically an atom causing an absorption line will be within that somewhat narrow line of sight so its blue or red shift is determined by all or most of the atom's velocity.

One crucial observation about the hydrogen emission line:

A quasar can give an electron to a proton traveling toward it at a velocity of multiples of c. The emission line from that new hydrogen atom is still observed here, though the light source can be moving away faster than c.
This single observation shows the velocity of the light source has no effect on the velocity of light propagation.

If the light source velocity had an effect then the radiation would have a negative velocity, which is meaningless and certainly not visible.

1.6 Sources Of Radiation

The possible single light sources:

1) An atom or molecule,

2) A solid (where molecular bonds hold molecules together in a cohesive lattice structure),

It releases energy as thermal radiation as it cools.

3) A liquid (where weak molecular bonds hold molecules to maintain fluid behaviors,

It releases energy as thermal radiation as it cools.

3) Condensed matter (where electromagnetic forces maintain a cohesive lattice structure; this lattice can result in a behavior of either a solid or liquid).

The photosphere of a star is condensed matter as a layer of liquid metallic hydrogen which cools by releasing energy as thermal radiation. This is how a star's surface temperature is measured.

The following celestial objects satisfy the above 3 criteria:

a) An atom or molecule (when it emits the energy from dropping to a lower energy state (this is a transfer of energy from one form to another),

b) A star will generate thermal radiation as it cools,

The distinction for the single source is crucial because a galaxy has many stars and does not have a surface enclosing all of them. Only each star has a cohesive light source with a surface and can exhibit the Doppler Effect,

Each celestial object radiates energy in all directions.

A star is a single light source so its motion affects its radiated spectrum by the Doppler Effect.

Note:
This behavior is used as one of several methods when searching for exoplanets.

A star and its planets orbit around the system's instantaneous center of gravity, called the barycenter.

In our line of sight, the star will be observed to cycle between moving toward Earth or away in our line of sight. As the planets orbit around the star, it gets pulled in a changing cycle based on the masses and distances of its collection of exoplanets. Our Sun is known to have this wobble. This technique for exoplanets is called either the wobble method or the radial velocity method.

Another exoplanets method is transit photometry, where an exoplanet's orbit causes a periodic dimming of the star in our line of sight. The period allows an orbit calculation, and the amount of dimming implies the size of the exoplanet.

(End Note)

A gas of atoms is not a single light source so a capture of light emitted from all its excited atoms will not exhibit the Doppler Effect. In a gas, each atom is an individual possible light source whose motion will affect its individual spherical radiated spectrum; this atom's generated spectrum is only the atom's emission line when it changed to a lower energy state.

The temperature of a gas is measured by the kinetic energy in the atoms or molecules in the sample.

An atom or molecule, whether as individuals or associated with others in a cloud, in the line of sight can cause an absorption line in the spectrum of a distant celestial light source.

The spectrum of a galaxy depends on the observer's line of sight. A galaxy has no single spherical surface generating its radiation.

A spiral galaxy has arms with dust clouds. The measured spectrum of a galaxy will vary from:

a) Directly above, sometimes called face on (this view is the brightest),

b) Directly on edge to the galactic disk where many stars are obscured in the field of view (this view is dimmest),

c) Any other angle.

Depending on the angle an elliptical type of galaxy can appear as either an ellipse or a sphere. This type covers both shapes.

Unlike the single light sources, a galaxy's spectrum depends on the observer's line of sight to the galaxy, where the collection of stars in the spectrum will vary.

As that radiation from all the stars in the galaxy propagates toward the observer, the light will pass through space where intervening atoms can reside. An atom in this path can cause an absorption line in the galaxy's observed spectrum.

A quasar has an AGN which does not exhibit the Doppler Effect. The AGN emits the flat spectrum of synchrotron radiation, as noted above. All of the many emission lines come from ions moving toward the AGN. Quasars are often dimmed by clouds of atoms which dim the AGN radiation. Any red shifts measured from those atoms do not reveal any motion of the quasar.

1.7 History of Red Shift

Excerpt from Wikipedia:

In 1912, Vesto Slipher measured the first Doppler shift of a "spiral nebula" (spiral nebula is the obsolete term for spiral galaxies), and soon discovered that almost all such nebulae were receding from Earth.
He did not grasp the cosmological implications of this fact, and indeed at the time it was highly controversial whether or not these nebulae were "island universes" outside our Milky Way.

Edwin Hubble is often incorrectly credited with discovering the redshift of galaxies.
These measurements and their significance were understood before 1917 by James Edward Keeler, Vesto Melvin Slipher, and William Wallace Campbell at other observatories.

Combining his own measurements of galaxy distances with Vesto Slipher's measurements of the redshifts associated with the galaxies, Hubble and Milton Humason discovered a rough proportionality of the objects' distances with their redshifts.

(Excerpt end)

Observation:

The critical conclusion was "a rough proportionality of the objects' distances with their redshifts" so this means the red shift is increasing by an increasing distance to the object. The connection to distance is not a velocity.

M31 galaxy and several others in its direction were found to have blue shifts, not red shifts, while both Magellanic Clouds, though rather near, have large red shifts.
These observations of both red and blue shifts just in our Local Group were not consistent.

Hubble recognized a problem with this data, noted in his 1936 book.

Excerpt from the 1999 paper titled: "The Local Group Of Galaxies" :

Hubble's (1936, p. 125) view that the Local Group (LG) is "a typical, small group of nebulae which is isolated in the general field" is confirmed by modern data. The zero-velocity surface, which separates the Local Group from the field that is expanding with the Hubble flow, has a radius Ro = 1.18 +/- 0.15 Mpc.

(Excerpt end)

Observation:

In Hubble's 1936 book, he proposed this solution for this problem: treat the Local Group as unique. This essentially avoids the inconsistency of red shifts and blue shifts. According to this paper, astronomers have "confirmed [through 1999] this 'zero-velocity surface." Rather than questioning Hubble's assumptions, the problem remains unresolved. The problem with absorption lines with distant galaxies should have been resolved around 1936.

The critical statement from this paper:

"Hubble discovered a rough proportionality between redshift of an object and its distance."

Observation:
Hubble recognized the hydrogen absorption line red shift is proportional to distance; velocity is not involved.

The redshift is caused by the distance through hydrogen in the IGM to the galaxy. The redshift is not from the galaxy but from its distance.

Researchers later concluded hydrogen atoms in the intergalactic medium cause this red shift which varies by the density in the path and direction so the red shift is roughly distance related but with a varying ratio.

This red shift has nothing to do with velocity. It is a mistake to relate them.

Hubble found a mix of red and blue shifts in the Local Group so he placed them on a 'zero-velocity surface' for these inconsistent objects separate from the Hubble Flow. There are differences in the IGM affecting red shifts being measured here.

The set of galaxy spectra used by Hubble is not public.

In 1936, the universe expansion (or Hubble Flow) had a known problem but astronomers never addressed it, so the anomaly persists even today.

There is a 2014 study titled:

"The Most Luminous z= 9-10 Galaxy candidates yet found: The Luminosity Function, Cosmic Star-Formation Rate, And The First Mass Density Estimate At 500 Myr"

Excerpt from this paper about galaxies with z from 9 to 10:

The identification of LBGs in the epoch of reionization makes use of the almost complete absorption of UV photons shortward of the redshifted Ly-alpha line due to a high neutral hydrogen fraction in the inter-galactic medium.

(Excerpt end)

These galaxies are not moving at z from 9 to 10, because the extreme red shift is from the IGM, not the galaxy.

This study is consistent with Hubble's conclusion in 1936.

1.8 Public Galaxy Spectrum Data

There are very, very few public spectra despite the multitude of galaxies. Wikipedia is the outlet for most astronomical data. For each galaxy, a relative velocity is often included, but not always, with a value in km/s or a z value, and sometimes both. A distance is usually listed also, but not always. Wikipedia does not include the spectrogram.

It is impossible to know what in the galaxy's unknown spectrum resulted in this calculated velocity. It is also impossible to know how this distance was measured. There are 2 common methods:
1) A Cepheid was found in the galaxy and its luminosity dimming by distance enabled a calculation.

2) The distance was calculated by Hubble's Law.

Hubble's Constant (HC) remains uncertain, still having a range of values. The HC value used in this calculation determines this distance.

All galaxy spectra should be public for independent analysis. The method used for the distance should also be public.

Wikipedia does not identify whether a Cepheid was used or which HC value was used.

1.9 Blue Shift Mistake

There are very few public spectra despite the multitude of galaxies but NASA posted the spectrum for the M31 Andromeda galaxy in a page for educators. This spectrum exhibits a blue shift. As with other galaxies, only the absorption line(s) from the atom in the line of light is affected. The hydrogen atom has a single wavelength but calcium has a pair.

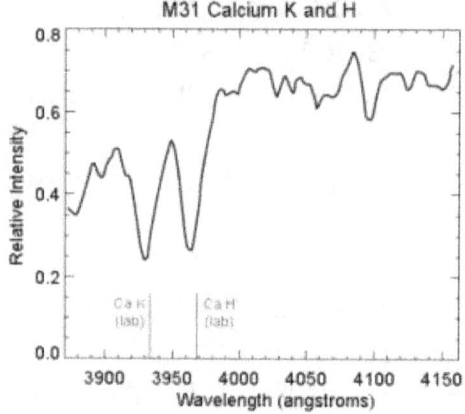

The crucial observation for a Doppler Effect is the entire spectrogram must shift together.
The NASA spectrogram is truncated at the ends. An astronomer in France posted his spectrograph for M31. If the page remains on Internet, it can be found by searching for "The Radial Velocity Measure of nearby galaxies" but this book does not copy his work. His post is interesting because it also includes the spectrogram from a type A1III star and this star also has the calcium absorption lines. This page contains a rarely seen stellar spectrogram.

For M31, only the calcium absorption lines are shifting so only the atoms are in motion not the galaxy. M31 has been assigned a relative velocity of -301 km/s as if it were approaching our Milky Way. This absorption line of a moving atom is not necessarily the velocity of M31. The correct method to measure the M31 relative velocity is explained later, in section 1.12.

It is a mistake to assign the velocity of an atom to the galaxy. Galaxies rarely have a public spectrum. Most galaxies will have single absorption line which is assumed to be from Hydrogen. Other elements have more than one electron so they have 2 or more lines depending on the element. There are a few, like M31, with lines from calcium. Since spectra are not public, it is impossible to know which galaxies have which element having its atoms in the line of sight.

1.10 Red Shift Mistake

This figure comes from a very frequently cited paper. This figure is in the paper found with a search for: "Dark Energy, and the Accelerating Universe: The Status of the Cosmological Parameters"

The text before this figure was: "This is one of the most dramatic examples of a macroscopic time dilation that you will get to see."

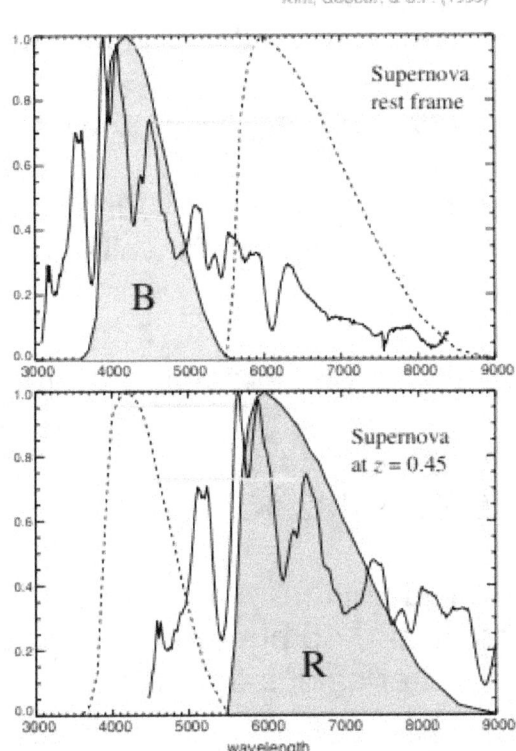

Figure 6: Slightly different parts of the supernova spectrum are observed through the "B filter" transmission function at low redshift (*upper panel*) and through the "R filter" transmission function at high redshift (*lower panel*). This small difference is accounted for by the "cross-filter K-correction" [9].

The crucial observation is the wavelength distributions are changing between B and R. They are definitely not just shifting to maintain the same distribution, as the Doppler Effect does. This figure is NOT a red shift by the Doppler Effect.

Each wavelength distribution is thermal radiation with the peak indicating a temperature so curve B appears hotter than curve R, with the B peak at a shorter wave length. When the peaks and patterns in thermal radiation are different between B and R, this comparison indicates different temperatures. The authors treat this difference in only the peak wave length as a time dilation.

This perceived dilation is wrong. Curve B has a peak wave length at about 4000 Angstroms while curve R has a peak at about 6000.

For reference, with those estimated wave lengths:

B peaking at 4000 is 7244K.
R peaking at 6000 is 4829K.

The calculated red shift from 4000 to 6000 is $z = 0.5$
The figure shows $z=0.45$ but the exact wavelengths are not provided for a better calculation here.

These spectrogram curves represent a blackbody temperature curve of an object in thermal equilibrium. Each actual calculated temperature from the figure is irrelevant because each value is not part of the paper's analysis.

The figure shows the surface temperature of the star is changing. This change in temperature is NOT a velocity of $z = 0.45$.

Observation:
Figure 6 shows in the spectrogram the temperature changing with a variable star's luminosity. This figure exhibits the spectrum of a star like a Cepheid variable star described earlier (in 1.2.j). The temperature change accompanies the brightness change.

This star is not a supernova; this is just a pulsating variable star whose temperature change causes a periodic luminosity curve, driven by its changing thermal radiation from its photosphere. Some of he thermal radiation frequency range is in our visual range, like from our Sun. This study used a collection of variable stars.

This is critical for this paper's conclusions.
There is no time dilation here. Figure 5 in the paper showed the changes in luminosity of the variable stars in the study.
Excerpt:

"this calibration procedure results in the remarkably tight distribution of light curves shown in the lower panel of Fig. 5."

(Excerpt end)

Observation:

These variable stars are not supernovae.
Conclusions based on those mistakes are wrong.

The authors mistakenly thought these variable stars were supernovae. There is a "tight distribution" because Cepheids exhibit similar luminosity curves.

A supernova has an extreme brightening. Because it probably involves ejection of material from the star, the dimming is unlikely to follow a repeatable pattern among these rare events.

It is beyond the scope of this book to describe other details about this particular paper. Among them: a) the other mistakes in this paper, with one mistake clearly revealed in only figures 5 and 6, and b) the consequences for the detailed conclusion justifying the claimed expanding universe. This often cited paper, having these errors, did not verify the accelerating expansion of the universe, as claimed.

1.11 Cosmological Red Shift

Cosmological redshift is the result of failing to fix the mistakes with red shifts.

Instead of addressing the mistakes with absorption and emission lines, the explanation of those red shifts is: the fabric of space is expanding, pulling apart the most distant objects to make their extreme velocity in their redshift a result of light distortion while propagating through the expanding fabric of space.

Excerpt from SAO Encyclopedia of Astronomy:

Laboratory experiments here on Earth have determined that each element in the periodic table emits photons only at certain wavelengths (determined by the excitation state of the atoms). These photons are manifest as either emission or absorption lines in the spectrum of an astronomical object, and by measuring the position of these spectral lines, we can determine which elements are present in the object itself or along the line of sight.

However, when astronomers perform this analysis, they note that for most astronomical objects, the observed spectral lines are all shifted to longer (redder) wavelengths. This is known as 'cosmological redshift' (or more commonly just 'redshift').

(Excerpt end)

Observation:

In a cosmological redshift, the wavelength at which the radiation is originally emitted is lengthened as it travels through (expanding) space. Cosmological redshift results from the expansion of space itself and not from the motion of an individual body.

A quasar has a hydrogen atom emission line with redshift but that is from the atom's motion and should never be used as the quasar's velocity. The SAO description ignores this mistake.
A galaxy spectrum has no emission lines, but only absorption lines from ions near the galactic corona or hydrogen in the intergalactic medium. This absorption line should never be used for the galaxy's velocity.

The SAO description combines the two separate mistakes to justify the simplified description.

Cosmological redshift violates the conservation of energy.

The reason:

The Doppler Effect at the moment of emission or absorption does not gain or lose energy. The Doppler Effect is either a transfer of energy or a change in its distribution within the sphere radiating from the source.

A blue shift or red shift at any other time is a change in the radiation energy with no identified partner for a transfer. This is a violation of conservation of energy because the energy transfer is undefined so the energy is not conserved. A red shift is loss of energy. A blue shift is a gain of energy.

By observation (noted above), the density of hydrogen atoms within the IGM and in the line of sight will affect the hydrogen absorption line shift.
This redshift behavior (from only the line of sight), unrelated to velocity, is one of the main reasons for claiming the false cosmological redshift.
Explanations of this change in the propagation of light from expanding space are cryptic.

Excerpt from Wikipedia:

The red shifts of galaxies include both a component related to recessional velocity from expansion of the universe, and a component related to peculiar motion (Doppler shift).The redshift due to expansion of the universe depends upon the recessional velocity in a fashion determined by the cosmological model chosen to describe the expansion of the universe, which is very different from how Doppler red shift depends upon local velocity. Describing the cosmological expansion origin of redshift, cosmologist Edward Robert Harrison said, "Light leaves a galaxy, which is stationary in its local region of space, and is eventually received by observers who are stationary in their own local region of space. Between the galaxy and the observer, light travels through vast regions of expanding space.

As a result, all wavelengths of the light are stretched by the expansion of space. It is as simple as that..." Steven Weinberg clarified, "The increase of wavelength from emission to absorption of light does not depend on the rate of change of a(t) [here a(t) is the Robertson–Walker scale factor] at the times of emission or absorption, but on the increase of a(t) in the whole period from emission to absorption."

Popular literature often uses the expression "Doppler redshift" instead of "cosmological redshift" to describe the redshift of galaxies dominated by the expansion of spacetime, but the cosmological redshift is not found using the relativistic Doppler equation which is instead characterized by special relativity; thus v > c is impossible while, in contrast, v < c is possible for cosmological redshifts because the space which separates the objects (for example, a quasar from the Earth) can expand faster than the speed of light. More mathematically, the viewpoint that "distant galaxies are receding" and the viewpoint that "the space between galaxies is expanding" are related by changing coordinate systems. Expressing this precisely requires working with the mathematics of the Friedmann–Robertson–Walker metric.

If the universe were contracting instead of expanding, we would see distant galaxies blue shifted by an amount proportional to their distance instead of red shifted.
(Excerpt end)

Observation:
The cosmological redshift comes from assumption in the cosmological model and "precisely working with the mathematics of the Friedmann–Robertson–Walker metric."

It is odd for someone to say
"It is as simple as that..." when this calculation is not simple as it is based on so many unfounded assumptions, including the cosmological model. The universe expansion arose from mistakes with red shifts. The model accepts that mistake.

The lack of communication among cosmologists has been demonstrated by relevant studies.

Recent studies of galaxies with high red shifts concluded the red shifts were from hydrogen in the intergalactic medium, not from a galaxy velocity.

When those studies are unknown or ignored, then the cosmological redshift is still proposed for the extreme red shifts wrongly assumed to remain unexplained.

1.12 Verifying Actual Velocities

A red shift is an attempt to at measuring an object's velocity in the observer's line of sight.
Currently, a red shift in a galaxy or quasar spectrum is treated as the 3-dimensional velocity for that object.
However the Doppler Effect is measured only in the line of sight so it is only a linear velocity always having a single vector in line with the Earth but either toward (blue) or away (red). That is wrong to just assume all those objects have no transverse velocity just because there was no attempt to measure it.

If cosmologists wish to continue using each red shift as a 3-D velocity then evidence is required for all those objects having no measured transverse velocity.

Actually, a sound foundation should be required for theories regarding the observable universe.

The correct 3-dimensional velocities are required to justify any theories involving objects beyond our Local Group. There are several, described in Section 4 (other concerns)

The calculation of 3-D motion requires precise position measurements at recorded times.

The change in position is a distance.

The change in position from start to finish defines the 3-D velocity vector.

The change in time from start to finish defines the motion's elapsed time.

The distance divided by the elapsed time results in the object's velocity, which has the observed 3-D vector.

Johannes Kepler developed his Laws of Planetary Motion using the precise celestial measurements recorded by Tycho Brahe. Kepler defined the motion of an ellipse which is 3-D motion. The ellipse could be at an angle to the observer.

Planets are millions of km or miles away but their motions can be measured by noting changes in position over a number of years. Using Earth's orbit as a base line, distance measurements are done by parallax.
This process becomes less reliable beyond the solar system because of relative distances in the 2 sides of a triangle.

Excerpt from Wikipedia:

Once a star's parallax is known, its distance from Earth can be computed trigonometrically. But the more distant an object is, the smaller its parallax. Even with 21st-century techniques in astrometry, the limits of accurate measurement make distances farther away than about 100 parsecs (or roughly 300 light years) too approximate to be useful when obtained by this technique. This limits the applicability of parallax as a measurement of distance to objects that are relatively close on a galactic scale. Other techniques, such as standard candles and spectral red-shift, are required to measure the distance of more remote objects.

(Excerpt end)

Unfortunately, that "spectral red-shift" method is unreliable, as noted in previous sections.

Standard candles are mentioned. They were used to calculate the distance to M31, or Andromeda Galaxy, for the first measurement of a galaxy beyond our Milky Way.
M31 is our closest galaxy, ignoring the Milky Way satellites.

The Cepheids in M31 can be used to calculate a 3-D velocity for M31.

Each Cepheid provides both a position and distance.

Making measurements over many years the star's 3-D motion and velocity can be calculated.

For example, if M31 were moving at $z=1$ then the Cepheid would change its position 1 light-year each year.

The distance to M31, using Cepheids, is 2.54 million light-years.
M31 and its stars are beyond the range of observing parallax.

The star's position change year to year will be difficult to resolve.

At that distance and at whatever velocity, everyone should agree any motion by M31 will require many human life-times to observe and measure.

With increasing distance, object positions lose precision.

Until these measurements have been made for a long enough time, there is no evidence for any motion by any distant galaxy.

Cepheids have a limited range.

The universe is huge and we can plot positions of many distant galaxies on the celestial sphere. Their velocities and distances are uncertain.

The reality is on the galactic distance scale they are essentially not moving. There is no evidence for any motion.

That simple statement falsifies the claims of the universe expansion and the big bang.

Our Milky Way is huge so stars must be observed and measured for years to measure their motion. The Gaia probe did that exercise for over a billion objects.

We recently learned about the motion of many stars in our Milky Way.

The universe beyond our Milky Way lacks that data.

Cosmologists must stop believing we have it.

Simulations have been developed using only unidirectional red shifts, while lacking any data for transverse motion.

Astronomers need a little humility. We really can't measure any motion of distant galaxies or quasars. They are only where we observe them in our lifetime. There is still wonder in their special distribution.

1.13 Varying Hubble's Constant

Hubble's Constant (HC) value varies among objects listed in Wikipedia when having both a public distance and velocity. The H value is easily calculated with V / D. A small random sample reveals different HC values are being used. A different H selection means each object's calculated distance is not consistent with others.

Astronomers claim Hubble's Law can translate a red shift into a distance, but it appears astronomers use varying values of Hubble's Constant for different objects to get a desired distance.

This constant converts a velocity (in km/s) into a distance (Mpc), where Distance = Velocity / HC

According to Wikipedia, the Hubble constant value is '"about 70 (km/s)/Mpc."

Here are several objects with their red shift velocity and the public distance (data from Wikipedia), with the corresponding H value that was used for that public combination.
Sometimes Wikipedia has both Mly and Mpc or just one.

The exercise is:

Calculate the HC from the D and V values provided.

For the following galaxies:
Z = red shift; V = km/s, D= Mpc

Public values, Z and D, for M104:

M104 Z = 0.003416 V = 1024 at D = 9.55 so from V/D HC = 107.23

M104 used HC higher than constant 70.

Public values, Z and D, for M60:

M60 Z = 0.003726 V = 1108 at D = 17.8 so from V/D HC = 62.25

M60 used HC lower than the constant 70.

Public values, V and D, for Hoag's Object:

Hoag's Object V = 12740 at D = 187.9 so from V/D HC = 67.8

Hoag's Object used HC lower than constant 70.

Public values, Z and D, for Abell 133:
Abell 133 Z = 0.00566 V = 16968 at D = 234 so from V/D HC = 72.51

Abell 133 used HC higher than constant 70.

Public values, Z & D, for 3C 273:

3C 273 Z = 0.158339 so V = 47501.7 at D = 749 so from V/D HC = 63.42

3C 273 used HC lower than constant 70.

Public values, Z & D, for NGC 67:

NGC 67 Z = 0.020734 so V = 6220 at D = 84.3 so from V/D HC = 73.77

NGC 67 used HC higher than constant 70.

A sample of 6 found none to use 70, the constant value for HC. Even if Wikipedia has the wrong value for HC, these 6 in the sample do not use the same HC value.

There are concerted efforts using different techniques to take many distance measurements to find the most accurate value of HC.
This small sample reveals that public HC value is not consistently used.

A sample of 6 might be small but all used different HC values.

Using different values of HC obviously affects the distance.

This simple exercise demonstrates Hubble's Constant can vary between objects. This sample had galaxies and a quasar.

Anyone can do the simple V/D calculation to get the HC being used for that object.

In this sample, one HC value was not used consistently.

Perhaps this observation is trivial but one should expect consistency or the values become random, when using a changing scale.

Both a galaxy spectrogram and the HC value being used are not public.

When a Cepheid was used not HC, that basis is unknown, though important.

1.14 Restoring Hubble's Law

A galaxy's red shift increases by distance due to hydrogen atoms in the IGM.
This ratio resulted in the original Hubble's Law.
Initially, this enabled a rough distance measurement. Then confusion arose when the red shift was treated as a velocity. That was a mistake because galaxies were assumed to be moving faster with distance, an illogical combination.

The preceding parts of this Section 1 Red Shifts correctly conclude a galaxy red shift must be ignored as a possible velocity.

Astronomers can still salvage something useful with the recorded galaxy spectra, which are never public. They must be archived somewhere, probably with each astronomer.

This is the current Hubble's Law equation, using HC for Hubble's Constant:

distance (in Mpc) = velocity (in km/s) / HC (in km/s/Mpc)

Cosmologists convened in July 2019 to discuss the Hubble constant crisis because different studies get different values. For a long time, it was 70. The meeting confronted a range of 69.8 to 72.4 but they failed to agree and the crisis continues.

It must be noted that Hubble used 46 galaxies for his first estimate and the result was 500 km/s/Mpc.

Even if it ranges widely, it could be somewhat useful as a very rough distance metric. "Something is better than nothing."

The revised equation must use the measured red shift value, NOT a velocity.

The new, appropriate equation, with z, not a velocity, becomes:

distance (in Mpc) = z / HCn

where HNn is HCnew in units of z / Mpc. (a new HC)

Example using Abell 133:
16968 km/s is z = 0.00566, D= 234 Mpc

The astronomer must have used a HC value of 72.51 km/s / Mpc to obtain that distance.

For the new equation:
HCn = 72.51 / (3×10^5) having no units in the numerator for the calculated value of z and the denominator units are still in Mpc.
The Abell 133 distance from 0.00566 / HCn is 234 Mpc

For this new equation to be useful, astronomers must recognize HCn varies around the universe because the IGM is not uniform. That is why attempts to precisely measure HC will continue to get different results in different directions.

The big problem in using this new distance equation is the negligence by astronomers in publishing the basis for a velocity or distance value. The previous topic 1.13 revealed an inconsistent "constant" value being used.

Every galaxy should have its spectrum public.

If that task is too difficult, then:

Every galaxy whose distance was calculated using a Cepheid must be identified that way.

The simple justification is: as astronomers change HCn over time or for different regions in space, a galaxy distance by a Cepheid does not change.

Whenever a Cepheid is found, then it reveals the HCn for that particular galaxy. Astronomers could be using their data productively but fail to do so.

Other names for HCn are probably better, like
IMF for "Inter-Galactic Medium Factor"

A name selection requires agreement by the community.

By not making the complete results public, it is impossible for others to make use of all observational data.

If astronomers managed data from Cepheids better, the distances calculated for distant galaxies could have a better approximation, by using a value adjusted by Cepheids, rather than the current "constant" somewhat in use now.

HCn will vary among different galaxy clusters. That is probably the cause of red shift anomalies like the Great Attractor and the Virgo Centric Flow.

The new equation's rules:

The galaxy must have the hydrogen absorption line red shifted. This line shift is caused by the IGM.

Some galaxies have calcium lines, with M31 as the notable example. The calcium ions are not in the IGM and do not exhibit a cumulative effect. Only hydrogen has the cumulative red shift behavior. That enables its use in this equation.

The equation for a galaxy cannot apply to the calcium lines. The calcium lines are also observed to have either a red or blue shift, but neither shift is cumulative.

If there are galaxies having absorption lines from other than hydrogen. Those atoms do not apply to this rule for only the hydrogen absorption line caused by the IGM.

Without public galaxy spectra, any such galaxies are unknown.

Emission lines cannot be used.

Quasars cannot be used, for any version of this law. This equation applies to only galaxies.

Quasars have emission line red shifts which do not come from the IGM.

Though the distribution of hydrogen in the IGM is not uniform, the hydrogen absorption line is a rough measurement of the variable quantity. Therefore, it can be used with a known but undefined error margin. Even a high error like 50% offers a rough distance suitable for a rough analysis of relative distances. Nothing can be precise on the galactic scale when using the IGM, which has never been sampled.

Here is an example scenario for this new equation:

A Cepheid is found in Abell 133 and its distance is now calculated at 235 Mpc, not 234, but its z value remains the same at 0.00566.

Therefore, the updated IMF = 0.00566 / 235 for the IGM effect toward this galaxy.
It is an astronomer's judgment whether to use this local value for other galaxies in its vicinity. The IMF in use for each galaxy should be public.

This is just a recommendation to apply the galaxy observational data for something useful. Galaxies cannot exhibit a velocity, but a rough distance calculation is possible.

Really, this is not a new equation. This is really the original equation with the original units for the red shift value in the equation.

Hubble observed a ratio between red shift and distance.

This equation restores that original equation

However, this equation also offers a new application to calculate the IGM effect to galaxies with Cepheids.

As a result, Hubble's Law gets its original equation restored. The description for Hubble's Law must eliminate the reference to a velocity.

Note:

The author uses IMF or HCnew or HCn for this new variable with the specific units of z / Mpc.

If cosmologists accept this new equation for Hubble's Law, the units for the current constant must be discarded.

In that case, the current variable named Hubble's Constant should be renamed to a name which reflects its use of a "not constant" value which changes among different galaxies.

1.15 New Red Shift Rules

The 2 rules are simple.

1) The red shift for a galaxy must be ignored as a velocity.

With a change in Hubble's Law and constant, the red shift can be used as an indicator for the IGM. The absorption lines are from atoms between us and the galaxy in the line of sight or in the IGM and are not from the galaxy's velocity.

2) The red shift for a quasar must be ignored. All red shifted emission lines are from the ions or protons moving toward the AGN, and are not from the quasar's velocity.

The term "red shift" has been used as if all red shifts are the same. They are not, so confusion results.

It is crucial to recognize there are 4 different red shifts.

A metallic element is one which is not hydrogen or helium.

The 4 distinct red shifts:

1) galaxy – hydrogen

2) galaxy – metal

3) quasar – hydrogen

4) quasar – metal

(1) the hydrogen absorption line is driven by hydrogen in the inter-galactic medium. This line is not from the galaxy.

(2) the calcium ion absorption line is driven by calcium ions near the galactic corona, in the case of M31 and others.

Sometimes, these lines can be blue shifted, like with M31.

(3) The high red shift comes from the Lyman-alpha emission line.

(4) The low red shift comes from metallic ion emission lines.

1.16 Conclusion of Red Shifts

The units for Hubble's Constant are restored to the original units from the 1920's, but with a new name. The Hubble's Law description must change to include the change in units.

The law relates a red shift to a distance. There is no velocity in this relationship.

This change resolves the current crisis in cosmology caused by the ongoing disagreement with the correct value of Hubble's Constant which is not constant for all galaxies.

A red shift can **never** represent the verified 3-dimensional velocity of an object.

The mistake with red shifts affects many theories involving galaxies and quasars. The most prominent impact is the universe expansion and its accompanying dark energy. Both are wrong, though both are widely accepted.

The false expansion is the reason for the big bang, so the big bang is wrong.

There is a big problem with the lack of public data. Claims are made of stars and galaxies but with no public data to enable their verification.

A simple example is the spectrogram for a Cepheid is not public to determine the correct explanation for its luminosity curve. There is no explicit list of which galaxies had their distance calculated by a Cepheid. That is important but such details are not public.

Cosmology lacks accountability for its claims.

Important note about X-rays:

X-rays are a problem for cosmology.
Thermal radiation typically spans from infrared to ultraviolet frequencies. A star's thermal radiation, seen as light, is within this frequency range.

Most emission lines from atoms range from visible to ultraviolet frequencies. As a general rule, any frequencies measured outside of this range of thermal radiation, like radio at the low end, and X-ray or gamma ray at the high end, were emitted by a source of synchrotron radiation. Wherever X-rays are observed, the explanation is wrong when proposing an extreme impossible temperature, like many millions of degrees.
A black hole violates this general rule because the hot accretion disk is claimed to emit X-rays but that requires an impossible temperature.

The spectrum of the claimed black hole reveals a source of synchrotron radiation. It is never thermal.

2 Gravitational Waves

This section 2 describes the detection of a gravitational wave.
The wave of an earth tide is also described.

2.1 Introduction to Gravitational Waves

Detecting a gravitational wave (GW) must be done indirectly because the GW has no real definition enabling a direct measurement.

LIGO is the Laser Interferometer Gravitational-Wave Observatory which was designed to detect these theoretical gravitational waves.

LIGO claims each GW detection, announced to the public, is from an astrophysical source, describing a binary of large masses (either is a black hole or neutron star) which spiral, collide, merge, and form one black hole at the end.

LIGO will be described to explain its method of detection.

2.2 Gravitational wave Origin

The origin of the theoretical gravitational wave might be trivia but here is its short story.

Excerpt from Wikipedia:

In 1905, Henri Poincaré proposed gravitational waves, emanating from a body and propagating at the speed of light, as being required by the Lorentz transformations and suggested that, in analogy to an accelerating electrical charge producing electromagnetic waves, accelerated masses in a relativistic field theory of gravity should produce gravitational waves. When Einstein published his general theory of relativity in 1915, he was skeptical of Poincaré's idea since the theory implied there were no "gravitational dipoles".
Nonetheless, he still pursued the idea and based on various approximations came to the conclusion there must, in fact, be three types of gravitational waves (dubbed longitudinal-longitudinal, transverse-longitudinal, and transverse-transverse by Hermann Weyl).

However, the nature of Einstein's approximations led many (including Einstein himself) to doubt the result.
In 1922, Arthur Eddington showed that two of Einstein's types of waves were artifacts of the coordinate system he used, and could be made to propagate at any speed by choosing appropriate coordinates, leading Eddington to jest that they "propagate at the speed of thought". This also cast doubt on the physicality of the third (transverse-transverse) type that Eddington showed always propagate at the speed of light regardless of coordinate system.
In 1936, Einstein and Nathan Rosen submitted a paper to Physical Review in which they claimed gravitational waves could not exist in the full general theory of relativity because any such solution of the field equations would have a singularity. The journal sent their manuscript to be reviewed by Howard P. Robertson, who anonymously reported that the singularities in question were simply the harmless coordinate singularities of the employed cylindrical coordinates.
Einstein, who was unfamiliar with the concept of peer review, angrily withdrew the manuscript, never to publish in Physical Review again.

Nonetheless, his assistant Leopold Infeld, who had been in contact with Robertson, convinced Einstein that the criticism was correct, and the paper was rewritten with the opposite conclusion and published elsewhere.

In 1956, Felix Pirani remedied the confusion caused by the use of various coordinate systems by rephrasing the gravitational waves in terms of the manifestly observable Riemann curvature tensor. At the time this work was mostly ignored because the community was focused on a different question: whether gravitational waves could transmit energy. This matter was settled by a thought experiment proposed by Richard Feynman during the first "GR" conference at Chapel Hill in 1957. In short, his argument known as the "sticky bead argument" notes that if one takes a rod with beads then the effect of a passing gravitational wave would be to move the beads along the rod; friction would then produce heat, implying that the passing wave had done work. Shortly after, Hermann Bondi, a former gravitational wave skeptic, published a detailed version of the "sticky bead argument".

(Excerpt end)

Observation:

From the beginning in 1922, gravitational waves were in doubt. Einstein himself tried to publish a paper denying them but withdrew that paper after being convinced his conclusion was wrong.

The account is not clear whether Einstein or Infeld wrote the final paper bringing the gravitational wave back to relativity.

2.3 Gravitational wave definition

Gravitational waves have a poor definition in terms of classical physics.

An excerpt from "NASA Space Place" which is simple but other public sites offer little or nothing in useful terms:

Gravitational waves are invisible. However, they are incredibly fast. They travel at the speed of light (186,000 miles per second).

Gravitational waves squeeze and stretch anything in their path as they pass by.

(Excerpt end)

Excerpt from the LIGO answer to "What are Gravitational Waves"

Gravitational waves are 'ripples' in space-time caused by some of the most violent and energetic processes in the Universe.
Albert Einstein predicted the existence of gravitational waves in 1916 in his general theory of relativity.
 Einstein's mathematics showed that massive accelerating objects (such as neutron stars or black holes orbiting each other) would disrupt space-time in such a way that 'waves' of distorted space would radiate from the source (like the movement of waves away from a stone thrown into a pond).

Furthermore, these ripples would travel at the speed of light through the Universe, carrying with them information about their cataclysmic origins, as well as clues to the nature of gravity itself.

(Excerpt end)

Observation to the LIGO definition:

The definition by LIGO has no details to enable the construction of a device for a direct detection and measurement of this gravitational wave. This deviates from classical physics where gravity is a measurable force between 2 known masses.

The wave definition does not define:

    a) the mechanism of its propagation, such as either longitudinal or transverse; it is certainly not electromagnetic radiation, or

    b) The medium for this wave's propagation, or

    c) The velocity of propagation (assumed to be c with no justification).

Space-time is a 4-dimensional coordinate system defined by relativity for the moving observer and cannot be a medium for an undefined wave.

LIGO built a system to detect an undefined wave having no defined medium for its propagation. LIGO expects this wave will squeeze and stretch the Earth, affecting the globe at multiple locations.
The multiple LIGO locations allow a triangulation of the source based on this minimal wave definition of only 'squeeze and stretch' and an assumed velocity.

LIGO is designed to detect a gravitational wave by monitoring Earth's crust for a disturbance which is analyzed and compared to computer generated templates assumed to match the expected results for this theoretical gravitational wave passing through the Earth.

LIGO should have verified whether this system detected a gravitational wave by this indirect method when based on only unverified assumptions.
Just one test with an actual merger of two known bodies would have confirmed the system is working as designed. This test was never executed. Without that crucial test and verification, LIGO had no basis for its operations.
With no verification by an independent observation, any LIGO detection could have been a different wave like coming from a terrestrial source.

LIGO has never tested this system with a known gravitational wave to verify any of the assumptions.

Every LIGO GW detection has no independent confirmation, to verify a GW claim.

2.4 Earth tide wave definition

Excerpt from Wikipedia:

Earth tide is the displacement of the solid earth's surface caused by the gravity of the Moon and Sun.
Its main component has meter-level amplitude at periods of about 12 hours and longer.

(Excerpt end)

There are 5 types of earth tide events in the LIGO history as the coincidental terrestrial source: Full Moon, New Moon, PeriGee, PeriHelion, and Moon-Jupiter alignment.
These 5 events will be referenced by a two-letter abbreviation:
FM, NM, PG, PH, MJ.

The Moon-Jupiter alignment event was a unique close celestial alignment with them and the Earth (in the solar system space they were far apart) on April 23, 2017.

The other 4 earth tide event types are well known to astronomers, needing no description here.

Though the MJ event happened only once, it is associated with 2 gravitational wave detections by LIGO, so MJ is in this list.
There is a frequent correlation between LIGO gravitational wave detections and earth tide waves

This correlation will be noticeable in the historical data in section 2.6. That observation alone is not sufficient for claims of causality.

2.5   Verification of the Terrestrial Source

The correlation between all LIGO gravitational wave detections and a terrestrial source might not be convincing evidence when presented alone.
However predicting a detection for a future wave detection is a confirmation of causality. The astrophysical source should be random in the universe. A prediction can be made based on this known, predictable terrestrial source. Having that prediction confirmed by a LIGO gravitational wave detection while the earth tide was present confirms the causality. This random merger event is between 2 unusual bodies. Neither has been directly observed as noted in Section 4 (other concerns).

The gravitational waves are claimed to originate at great distances in the universe. They should not be predictable.

a) Hypothesis Development

Historically LIGO reports detections within 2 days of an earth tide for more than half of the detections.
In observing run O3 there are usually additional detections outside of this narrower range.
In O1 and O2 9 of 11 were within 2 days; the other 2 were at 4 days.
In O3 with the increased sensitivity a small number of detections can be up to 7 or 8 days from that earth tide.
In O3 21 of the 41 merger detections were within 2 days.

The analysis reveals every earth tide event will always result in 1 or more LIGO wave detections.

More than half detections are within 2 days and there are usually a few more detections in the range of 3 to 5 days.

The hypothesis: LIGO will report gravitational wave detection for the ripple of an earth tide event.

In observing run O3, the sequence of one earth tide event triggering more than one detection is observed multiple times.

The peak of each earth tide is known and predictable and the Earth rotates once per day so the influence is not present only at the moment of the peak alignment.
The alignment of Earth, Moon, and Sun for a full moon or a new moon takes a number of days for its effect to begin and end.
O3 exhibits a wider range than O1 and O2 for the span of days around the earth tide events.
The Earth's crust is solid so the earth tide is different at the surface than on an ocean.
The ripple in the crust from an earth tide is not precisely predictable to a specific date for the LIGO detectors.
However one should expect its ripple to span beyond just the date of its peak.

b) Prediction Development

Because of LIGO's inherent inconsistency which is increasing in the course of O3, the prediction could not be limited to only an exact date so a range is required. A range should be restricted enough to provide a valid prediction for a valid test.

On November 9, I noticed a full moon coming on November 12 so I gave my prediction to LIGO on the morning of November 10.

On November 9 LIGO Scientific Collaboration public Facebook page had a post about their new November 9 detection.
I selected this post for my prediction in a comment.

The LIGO Facebook page allows comments from the public but not posts.

I intentionally made the prediction for several explicit ranges to avoid the easy dismissal of a '"One time lucky guess."

At the moment I made the prediction the last O3 detection was on November 9.

c) Prediction

The prediction was given to LIGO Scientific Collaboration at 10 am my time or 16:xx UTC:

(begin of text)

Predictions:
There will be LIGO detections between November 10 and 14, between November 21 and 25, between November 24 and 28.
There will be several other detections before and after these narrow ranges.
I was late with this prediction but detections were already reported on November 5 and 9.

Since LIGO began reporting detections it reports them in clumps with more in each clump in the O3 run (less in O1/O2).
For example in 2017 August 14, 17, 18 had detections.

(end of text)

d) Test Results

These are the gravitational wave detections by LIGO after the prediction.
The format for each line:

LIGO detection ID, with a brief comment

S191110x, at 18:09:05 UTC or 2 hours after prediction

S191110af, at 23:10:59 or 7 hours after prediction

S191117j, or 3 days after the first range

S191120a, for the second range of dates in the prediction

S191120at, also for the second range

S191124be, also for the second range

S191129u, for the third range of dates in the prediction

e) Summary of Results

There were 2 detections within 7 hours of the prediction's first range of dates.

Detection followed 7 days later.

The other two ranges were later in the month and also part of the prediction. Those 2 ranges of dates also had detections (4 of them) as predicted.

The prediction defined 3 ranges of dates for detections and all 3 predicted ranges had detections where 3 detections of the 6 were within 2 days which is the observed range for over half the detections.

Here is a comparison of the deviations between LIGO detections to the earth tides:

These 7 detections had these deviations in days from the triggering earth tide:
-2, -2, +5, -3, -3, +1, +3.

Each range had its clump of detections as expected in the prediction.

f) Conclusion from the Test Results

The prediction of wave detections within specific dates was confirmed by these LIGO detections and the hypothesis was validated by this simple test. Therefore:

LIGO declares gravitational wave detection for the ripple of an earth tide wave, making it possible to predict the wave detections.

The LIGO system is not consistent with its detections in its history as demonstrated by 2 detections on a single day being reported twice in this small sample of only 7 detections. This sample is not a random distribution, but it contains the same behavior as found in the history of GW detections.

The distribution of LIGO detections is driven by the periodic earth tides.

2.6 Historical data

This table presents the historical data for a convenient reference.

All of the LIGO gravitational wave detections are listedthrough March 2020, along with their associated earth tide events.

This table is in chronological order.

All the date entries use the same 6-digit date format of YYMMDD, Where the detection will have one or two letters before the date
and one or two letters after the date. The earth tide dates have two letters before the date. This abbreviation was described above.
2.2.
For example, PG150914 means periGee on 2015 September 14.

To distinguish between the 3 observing runs: 2015 was O1; 2017 was O2, and 2019 was O3; 2020 has continued as part of O3.

The table lists all the gravitational wave detections, with the closest earth tide event.

LIGO Event , Earth Tide

GW150914 , PG150914

GW151012 , NM151012

GW151226 , FM151225

GW170104 , PH170104

GW170608 , FM170609

GW170729 , PG170802

GW170809 , FM170807

GW140814 , PG170818

GW170817 , PG170818

GW170818 , PG170818

GW170823 , NM170821

LIGO Event , Earth Tide

S190405ar , NM190405

S190408an , NM190405

S190412m , PG190416

S190421ar , FM190419

S190425z , MJ190423

S190426c , MJ190423

S190503bf , NM190504

S190510g , PG190513

S190512at , PG190513

S190513bm , PG190513

S190517h , FM190518

S190518bb , FM190518

S190519bj , FM190518

S190521g , FM190518

S190521r , FM190518

LIGO Event , Earth Tide

S190602aq , NM190603

S190630ag , NM190702

S190701ah , NM190702

S190706ai , PG190705

S190707q , PG190705

S190718y , FM190716

S190720a , FM190716

S190727h , NM190731

S190728q , NM190731

S190808ae , PG190802

S190814bv , FM190815

S190816i , FM190815

S190822c , FM190815

LIGO Event , Earth Tide

S190828j , NM190830

S190828l , NM190830

S190829u , NM190830

S190901ap , NM190830

LIGO Event , Earth Tide

S190910d , FM190913

S191910h , FM190913

S190915ak , FM190913

S190923y , NM190928

S190924h , NM190928

S190928c , NM190928

S190930s , NM190928

S190930t , NM190928

LIGO Event	Earth Tide
S191105e	FM191112
S191109d	FM191112
S191110x	FM191112
S191110af	FM191112
S191117j	FM191112
S191120aj	PG191123
S191120at	PG191123
S191124be	PG191123
S191129u	NM191126
S191204t	FM191212
S191205ah	FM191212
S19121212q	FM191212
S191213g	FM191212
S191213ai	FM191212
S191215w	FM191212

LIGO Event , Earth Tide

S191216ap , PG191218

S191220af , PG191218

S191222n , NM191225

S191225aq , NM191225

S200105ae , PH200104

S200106au , PH200104

S200106av , PH200104

S200108v , FM200110

S200112r , PG200113

S200114f , PG200113

S200115j , PG200113

S200116ah , PG200113

S200128d , NM200124

S200129m , NM200124

S200208q , FM200209

S200213t , PG200210

LIGO Event	Earth Tide
S200219ac	NM200223
S200224ca	NM200223
S200225q	NM200223
S200302c	FM200309
S200303ba	FM200309
S200308e	FM200309
S200311bg	PG200310
S2003116bj	PG200310

Of the 90 detections through March 2020, these were the distribution of the days from an earth tide peak date:

0 days = 10 GW,

1d = 23,

2d = 22,

3d = 16, so 71 out of 90 were within 3 days of an earth tide date.
4d = 8,

5d = 3,

6d = 3,

7d = 4,

8d = 1

Clearly, there is a connection. However, LIGO is very sensitive so it is inconsistent. There were 38 earth tide events triggering 90 GW detections.

The following observation compares the deviation spread in the history of GW detections over the 3 observing runs.

Observing runs O1 and O2 had no GW beyond 4 days of an earth tide peak. Only one perigee resulted in 3 detections. After the upgrade in 2019 to begin run O3, the spread became wider.
With run O3, LIGO was not detecting more of the distant mergers as claimed but LIGO actually detected more gravitational waves from the same earth tide as the terrestrial source spanning more than one day.

For clarity all LIGO detections are listed. Beginning with O3, LIGO posts all their detections in the GRACEDB site, including those whose analysis failed to obtain the merger pair. As a result, some wave events recorded in GRACEDB are not posted in Wikipedia where probabilities are assigned to the various merger combinations. This analysis covers all LIGO GW detections.

2.7 Doubts of LIGO claims

First:

Page from April 11, 2018 titled:
"Danish Group's Doubts That LIGO Discovered Gravitational Waves Resurface"

Excerpt:

A group of physicists in Denmark, which doubted last year whether American experiments to detect gravitational waves had actually confused noise for signal, has reared its head once more. The *New Scientist* reported earlier this week that the group, from the Niels Bohr Institute in Copenhagen, independently analysed the experimental data and found the results to be an "illusion" instead of the actual thing.

Sabine Hossenfelder, a theoretical physicist at the Frankfurt Institute for Advanced Studies, wrote on her blog on November 1:

… the issue for me was that the collaboration didn't make an effort helping others to reproduce their analysis. They also did not put out an official response, indeed have not done so until today. I thought then – and still think – this is entirely inappropriate of a scientific collaboration. It has not improved my opinion that whenever I raised the issue LIGO folks would tell me they have better things to do.

(Excerpt end)

Second:

Physicist Sabine Hossenfelder mentioned LIGO in her 'backreaction" blog again in a post over a year later. Her blog entry on September 4, 2019 was titled: 'What's up with LIGO?'

Her post included a link to a .de web page for its news story.

Google Chrome did a translation for this web page.
Its title in English: "Fake news from the universe"
Others can either read the original or get their own translation.
Otherwise, the reader must decide whether the following translation is correct. The lack of an appropriate reaction suggests no grounds for LIGO against the story.

Excerpt from the browser's translation:

For two months now this new "window to the universe" is in operation and finds - nothing. Although there were not a few alerts from LIGO / VIRGO, but not a single signal that could have confirmed the large terrestrial or space telescopes. The astronomers are already slightly annoyed about the wasted observation time and ask questions. What's happening?

This surprising result should be a reason to take a closer look at the publications on gravitational wave observation over the last three years.

The statistical disturbances caused by random vibrations of the 3000 km distant LIGO laboratories had inexplicable correlations. Only the gravitational wave itself should be visible in both laboratories - with a corresponding delay due to the light propagation time. After ignoring the results of the Danish working group for a while, a group of eight scientists traveled to Copenhagen in August 2017 to discuss data analysis with their critics.

The gravitational wave researchers had to admit some mistakes, among other things, that the central figure in the journal Physical Review Letters was not created with the original data, but prepared for "illustrative purposes" - embarrassing for an article that was downloaded a hundred thousand times and was the basis of the Nobel Prize 2017, At the meeting in Copenhagen the photo of the blackboard was created. One of the leading LIGO scientists, Duncan Brown, promised to work with his colleagues for the correction - which has not happened to this day.

Meanwhile, Jackson's group has even proved that a so-called template, a theoretically calculated signal used for analysis, was subsequently replaced.

It is extremely remarkable that with this unprejudiced method none of the more than twenty detected gravitational wave signals could be reliably detected - except for the first signal GW150914 in September 2015. Now one could argue that this first signal provided proof and danger banned that the following signals were caused by arbitrary filtering of random noise.

Of course, this is still no evidence of manipulation, but it would be given the quite existing internal doubts certainly appropriate that LIGO makes its own investigations to more transparent.

However one evaluates these events, it remains the fact that after three more years of operation and meanwhile triple sensitivity of detectors GW150914 is still the strongest signal of all. A coincidence that gets stranger every day.

For many, therefore, the strongest evidence for gravitational waves is based on the August 2017 GW170817 signal discovered by LIGO and then confirmed by the Fermi (NASA) and Integral (ESA) gamma-ray / gamma-ray telescopes, but with very weak signal. at any rate, it was presented at the press conference.

In truth, it was the other way round: Fermi had sent the notification email first, and LIGO needed four hours to "predict" the sky position - which was consistent with the coordinates already known. The false impression that LIGO was the first one arose simply from the fact that after an explicit request by LIGO the subject line of the alert mail had been modified (see picture).

In addition to these inconsistencies, well-known experts contradict the interpretation that the signal comes from merging neutron stars. According to an author collet from nine renowned institutes, this is only possible through "extreme models" of the corresponding galaxies, while an Italian working group assigns the gamma-ray signal (or the afterglow) to a fusion of white dwarfs. But they can not send gravitational waves.

So there remain considerable doubts as to whether GW170817 was really confirmed by other telescopes or whether it was even a gravitational wave.

(End of excerpt from the translation)

If this story is accurate, then it is truly a sensational revelation.
According to Sabine's posts, LIGO has not responded to these questions being asked of their claims.
Perhaps the reader will reach this disturbing conclusion:
This is not proper science and so LIGO has no credibility.

2.8 LIGO Detection Details

This section 2 explains how LIGO declares an astrophysical source for what was actually a terrestrial source.

The LIGO details for gravitational wave detections are a separate topic after the dates of claimed GW detections.

This section 2.8 addresses that merger description from LIGO.

A gravitational wave has no definition, for either the wave or a medium for its transmission.

LIGO developed a system that could detect any disturbance of the earth by placing the interferometers distributed at several locations around the globe.
With that system in place LIGO developed software to analyze the disturbances being detected and then determine the source of that presumed wave.
LIGO developed templates for possible merger scenarios with the expectation those templates could be found in this signal that was designed to capture a movement in the crust even smaller than a proton.

LIGO relies on this signal analysis, but this analysis has never been tested and verified.

The infamous chirp described by LIGO is not part of the detected wave. The earth tide does not have to mimic any aspect of the theoretical gravitational wave. The earth tide must only trigger the LIGO analysis which reacts to a disturbance. If the template is detected then the LIGO team can freely create its description for the event. The word chirp makes media interactions congenial.

LIGO's design magnifies any disturbance many times. LIGO is proud of the extreme sensitivity in its interferometers.

Excerpt from LIGO:

The longer the arms of an interferometer, the smaller the measurements they can make. And having to measure a change in distance 10,000 times smaller than a proton means that LIGO has to be larger and more sensitive than any interferometer ever before constructed.

(Excerpt end)

The highest probability for this unverified system is if LIGO can really identify a 'chirp' with any earth tide wave, that ringing is from the LIGO design not from the earth tide wave. Any ringing claimed by LIGO from an earth tide wave is the edge of this surface wave transitions at the detectors being extremely amplified by the system's design.

LIGO can report multiple gravitational waves around the peak of an earth tide wave with detections both near the start, at the peak, near the end, and on the days in between.

For example, the full moon on 12/12/2019 had 2 GWs before peak, 1 GW on the peak, and 3 GW after the peak. LIGO claims it found the chirp with a gravitational wave detection but since nothing in LIGO has been verified the LIGO gravitational wave descriptions are meaningless.

Nothing in the LIGO process has ever been verified. Each earth tide event triggers the LIGO analysis and an unverified description is provided.

Until LIGO actually verifies the details of any detection all those details are invalid, including the chirp.

LIGO's design magnifies any disturbance many times; LIGO is proud of this sensitivity.
If LIGO can really identify a `chirp' with any full moon or new moon passing overhead, that ringing is from the LIGO design not from the earth tide wave. LIGO using unverified software claims it found the chirp but with no independent observational evidence as verification of the actual merger event.

It is very difficult to grasp how the complexity of using tiny ripples detected at several widely spaced detectors can result in the very detailed conclusion:

1) the type of each body, either black hole or neutron star,
2) the precise mass of each body,
3) a roughly described coordinate in the sky (the margin of error is undefined),
4) a roughly described distance to the event (the margin of error is undefined)
5) the remaining mass after the merger,
6) the spin of this remaining mass.

This is truly a major accomplishment (awarded the 2017 Nobel Prize in Physics) when the entire system (hardware and software) was never tested with even one such merger to verify whether any of the details were correct. Using the word incredulous is just an opinion.

2.9 Conclusion of Gravitational Waves

Gravitational wave detection has **never** been verified to match its description,

Since nothing in the LIGO claims has been verified, both the GW detections and their merger descriptions must be ignored.

The LIGO false GW detections cannot be used as evidence for black holes and neutron stars.
LIGO's false GW detections have confirmed nothing about relativity, including its theoretical GW.

Cosmology lacks accountability for its claims.

3 Relativity

This section 3 describes the basics of Einstein's theory of general relativity and how it is applied to the context of the universe.

For example, black holes were described by both Einstein and Hawkings, and are claimed by many as being observed.

3.1 Introduction to Relativity

Some advocates of relativity believe in the adage: "Spacetime tells matter how to move; matter tells spacetime how to curve."

It is difficult to believe anyone well educated with the science of physics would believe that laughable statement. That possible reaction and its justification will be explained in section 3.9.

3.2 Time definition

Time is a measurement. Time is not motion but time is used to measure motion.
Chemical reactions are not instantaneous but we use time to measure them.
Chemistry has nothing to do with motion.

60 seconds in a minute probably came from counting a person's pulse.
"A normal resting heart rate for adults ranges from 60 to 100 beats per minute."

If anyone needs an approximate time measurement they can just count their pulses to count a number of seconds (though not exact).
Performing this count does not create time.

The current date and time are a reference with both incrementing at a defined rate. They describe my current now as part of the measurement of event sequences in the universe.
There is only now and it is impossible to pick a different time for now, except for an accepted change like the daylight savings time adjustment. Essentially we agree to change the time of noon or the Sun at its highest compared to the Sun rise. Earth's axial tilt results in changes in the time elapsed between sun rise and sun set. Our time of day does not begin with sun rise but the 24 hours in one rotation are roughly centered on the highest sun, with 12 before and after.

Changing the date and time for now does not mean my now has changed to a different now.

If someone says time travel is possible that is like saying one can change a watch and calendar and magically physically move to a new date and time as selected.

3.3 Describing A Position

The observer selects or defines the coordinate system based on the requirements for a measurement.

In a laboratory for an experiment, the simple Euclidean geometry is often used.
3 linear dimensions are defined with the desired scaling. 3 letters are often used for the 3 linear dimensions, with x for left/right, y for up/down, z for in/out. The scaling is also defined, such as mm or inches, or perhaps much longer increments.
The respective dimensions must be referenced to physical space.
One example is X0, Y0, Z0 is often at the lower left corner of the working space; alternately X0 could be defined at the middle of the working space. This definition allows someone else to repeat an experiment exactly, by recording the positions using the same coordinate system definitions.

Time is sometimes considered a 4th dimension. To measure motion, the time difference between position measurements enables the calculation of velocity (distance per unit of time).

For measuring positions on Earth, the respective observers can use the GPS coordinate system so measured positions can be shared and repeated. The GPS coordinate system is referenced to the center of the Earth.

For measuring positions in the universe, the respective observers can use the celestial coordinate system so measured positions can be shared and repeated.

Excerpt from Wikipedia:

A celestial coordinate system is a system for specifying positions of satellites, planets, stars, galaxies, and other celestial objects. Coordinate systems can specify an object's position in three-dimensional space or plot merely its direction on a celestial sphere, if the object's distance is unknown or trivial.

(Excerpt end)

Observation:

The celestial coordinate system uses two dimensions having angular values.
The celestial coordinate system is referenced to the center of the Earth but is offset by the observer's current position on Earth's surface and their local time; this transformation enables its consistency during Earth's rotation.

A third linear dimension for the distance to the object allows the complete description of an object's position anywhere in the observable universe, using Earth as the reference.

3.4 Types Of Motion

a) Commanded motion

The observer can define a coordinate system to measure the current location of any object including oneself.

After the observer's coordinate system is anchored to physical space, by using a reference point at a physical location for each dimension, the observer can use the dimensions for accurately describing an intended motion in relation to that reference point, a real position in real space.

b) Measured motion

In classical or Newtonian physics, objects are moving subject to external forces. None of these objects are in motion using a coordinate system. No motions are executed using coordinates for the destination like moving from one coordinate to X1.2, Y2.3, Z3.4. Every object is in either motion or stopped, based on the sum of all forces acting on it. Bodies in motion can have their position measured at intervals to calculate their velocity, from a measured distance over a measured time. The 3 fundamental forces of gravity, electric, and magnetic all act by the inverse square of distance so each force decreases as the distance increases.
Every force on a mass has a vector for the resulting acceleration. This force can be any of the 3 fundamental forces: gravity, electric, magnetic.

When knowing the mass, charge, and distance for each relevant body these forces can be calculated.

When knowing the force and mass, the acceleration can be calculated.

c) Coordinated motion

The gravitational slingshot trajectory of a space probe is an example of coordinated motion, or simultaneous motion.

When NASA calculates a trajectory of a space probe it uses the force of gravity as defined by Newton.

Universe Today had a 2014 story titled "How do gravitational slingshots work?"

The story describes how NASA calculates these slingshots to execute a change in a probe's trajectory around a moving planet. NASA has certainly demonstrated their technique with numerous successful missions.

The calculation of a slingshot involves these critical values:
a) the mass of the probe

b) the mass of the planet

c) the velocity of the probe

d) the velocity of the planet.

During the probe's approach there is the mutual force of gravity between the two bodies where the paths of both bodies are affected simultaneously. Obviously, the probe with a rather small mass is affected much more than the planet.

These calculations are based on Newton's gravity.
Our solar system has all the bodies simultaneously rotating around the instantaneous center of gravity, also called the barycenter.
Relativity is based on space-time curvature driven by a gravitational field. A gravitational field provides free fall acceleration toward that body which is spherical having uniform density. Conforming to those rules, this body exerts this field which can be calculated from the mass density and radius.
The mass of the observer, a smaller mass than the main body, is not involved in this free fall calculation.
The heavier body will not free fall toward the lighter body. On Earth, applying a force to a body to lift it gives the body potential energy. Upon releasing the body it will have free fall acceleration toward the heavier body. That free fall behavior is not part of this calculation for a space probe.

NASA never uses a gravitational field in its calculations for a slingshot trajectory. NASA does not use space-time curvature.

Relativity assumed gravity had a velocity limit of c. NASA assumes gravity is instantaneous, when using the Isaac Newton equations which have no variable for time.

The gravitational slingshot is confirmation of Newton's gravity.
The fact astronomers predicted and discovered the planet Neptune in 1846 verified Newton's force of gravity is correct.

The barycenter behavior in our solar system is also confirmation of Newton's gravity.

Space-time curvature requires commanded motion by the moving observer when the commanded path is curved.

During the barycenter coordinated motion, with no moving observer present, space-time is quite irrelevant.

That suggests relativity is the wrong theory for the evidence.

There has never been an observation where space-time correctly matches an observation when gravity did not.

There is no evidence for space-time while the force of gravity consistently has the evidence.

3.5 Travel Of Light

Visible light is part of the electromagnetic radiation spectrum being generated by synchronized, perpendicular electric and magnetic fields which propagate through a vacuum at the measured velocity called the constant c. These fields have a period of oscillation measured as either wavelength or frequency. The wavelength is often measured in nanometers. This propagation velocity can be reduced by the medium by a factor called the diffraction index.
Light is not a particle and has no mass so a change in the medium will cause an immediate change in the velocity. This behavior is observed with light bending at the surface of water in a glass.

A prism demonstrates light is inherently a wave because particles would not spread in a coordinated manner as observed; only a wave propagating in a medium matches the observation. Any behavior where light might appear as a particle is due to the circumstances of the observation.

Light will always propagate in a straight line unless the width of the beam propagates through a medium having changes in the medium across the beam's width, like in a prism.

Absorption and emission lines in a spectrum are not a particle behavior. They are either a measured wave length or frequency within the continuum of radiated energy. A continuum cannot be quantized.

A photon is just an abstraction of one wavelength of light.

Light can never exhibit the behavior of commanded motion. Light cannot follow the dimensions, or travel to a particular coordinate, within a defined coordinate system.

The force of gravity is the mutual force of attraction between two defined masses.

The propagation of light, which is not an entity having mass, is never affected by the force of gravity.

Space-time Definition

Relativity is a theory defined to be background independent.

Excerpt from Wikipedia:

Background independence is a condition in theoretical physics, that requires the defining equations of a theory to be independent of the actual shape of the spacetime and the value of various fields within the spacetime. In particular this means that it must be possible not to refer to a specific coordinate system—the theory must be coordinate-free. In addition, the different spacetime configurations (or backgrounds) should be obtained as different solutions of the underlying equations.
(Excerpt end)

Observation:

This "coordinate-free" basis is appropriate for the theory of relativity because relativity describes changes to only the moving (or non-inertial) observer's reference frame, or their 4-D space-time.

Relativity never uses or needs the background coordinates for the observer.

The underlying equations are also important because they demonstrate the context on the moving observer.

The moving observer gets the space-time curvature. Space-time is a coordinate system having 4 defined dimensions: ct, x, y, z.

This is the Euclidean geometry but with a 4th dimension ct allowing time to be introduced as a linear dimension (when multiplied by c) with units compatible with the other 3 having standard linear dimensions.

The commanded motion of the moving observer is being manipulated by the observer's motion in a gravitational field.

The combination of change-x, change-y, change-z, change-ct are used to calculate the space-time interval for the geometric transformation done in Einstein's equations.

This transformation involves only changes in the observer's position and never uses a reference to a physical location.

The differential inputs into the metric tensor are: cdt, dx, dy, dz.

These represent the change in the moving observer's position for each increment of time during that motion.

This incremental motion described using differentials is integrated over a period of time to get the path of motion.

This description is a simplification but sufficient, because it has the important details.

Relativity does not require a connection to a coordinate in physical space, when working solely within the context of the moving observer's reference frame and manipulating the path of motion within that context.

Isaac Newton is said to have worked in a background dependent context with absolute space and absolute time. In other words, objects could be described by their coordinates in physical space. Time is a measurement completely independent from a position in space,

The terms of geometry enable the definition of the observer's coordinate system, with its dimensions or axes and their scaling.
These dimensions are connected to physical space by relating each to a particular point in physical space. A simple example with the Euclidean geometry is defining coordinates $X0, Y0, Z0$ at the lower left corner of one's working space. The observer selects the scaling, such as inches or mm.

In this case, the coordinate system has become background dependent by defining its coordinates in physical space.

This is the context often used by an observer in a laboratory.

To become background independent the observer must make all measurements of changes in positions relative to the observer's location without ever needing to define an absolute position of any objects.

In Newtonian physics (i.e., not relativity) when used for cosmology, the context of background dependence is convenience. Each behavior in Newtonian physics is usually connected to a specific point in the physical space of the universe. The observer defines how the positions are measured when the reference point in space is selected. This reference point could be in a laboratory or in distant space

The forces of gravitational fields, electric fields, magnetic fields are all based in movable objects and are never required to be anchored to a point in physical space. The distance is crucial for these forces. Their position in physical space is not important. Gravity works the same on Earth, the Moon or any other place in physical space.

I Newton's equations were applied in a background independent context, there is no coordinate system to define the dimensions required to measure a position.

Without defined positions, only distance measurements are possible, so any position is relative to another. The 3 fundamental forces use only a distance, never a position. This context of only distances is possible, but impractical and essentially ineffective and certainly inefficient. The easiest distance calculation with 3 dimensions is the comparison between their respective coordinates in the defined coordinate system using the observer's scale for each dimension, using basic trigonometry.

In the GPS spherical coordinates, the radius of the earth is the linear distance for the hypotenuse of a triangle. The sine function with either the latitude or longitude angle enables the calculation of a distance on the surface for that angle, or on the surface by that radius. In practice, maps use scaling so one unit of measurement is proportional to an actual distance, like x inches per mile.

Instead of a Euclidean geometry or the GPS dimensions, a similar technique is used for the celestial coordinate system where the two planes are related to the fixed point at the center of the Earth. Observers around the world can adjust this coordinate system for their location relative to the center of the earth.

In the celestial spherical coordinates, the known linear distance from Earth is the hypotenuse of a triangle. The sine function with either the Declination or Right Ascension angle enables the calculation of a distance between objects for that angle.

This is the typical background dependent context for a coordinate system usable by anyone on Earth.

Relativity uses the concept of frame of reference.

Excerpt start:

The motion of a body can only be described relative to something else—other bodies, observers, or a set of space-time coordinates. These are called a frame of reference.

In physics, a frame of reference (or reference frame) consists of an abstract coordinate system and the set of physical reference points that uniquely fix (locate and orient) the coordinate system and standardize measurements.

(Excerpt end)

Observation:

This "abstract coordinate system" can be the observer's context, and in relativity it is.
Relativity implements its space-time curvature as changes in the observer's frame of reference.

Excerpt start:

General relativity generalizes special relativity and refines Newton's law of universal gravitation, providing a unified description of gravity as a geometric property of space and time, or spacetime. In particular, the curvature of spacetime is directly related to the energy and momentum of whatever matter and radiation are present. The relation is specified by the Einstein field equations, a system of partial differential equations.
(Excerpt end)

Observation:
A statement above is misleading because it omitted critical words. It should be fixed like this with the added text in < > :

In particular, the curvature of < the observer's> spacetime is directly related to the energy and momentum of whatever matter and radiation are present < at the observer >.

This distinction is very important.

Curvature is NOT related to whatever matter and radiation are present ANYWHERE. The curvature is directly related to the moving observer. This is just semantics but it should be correct and clear.

Relativistic behaviors affecting the observer's space-time do not apply to the physical universe when limited to the observer's reference frame.

However, cosmologists consider space-time as a real thing, which is clearly a mistake.

Excerpt from Wikipedia:

The shape of the universe is the local and global geometry of the universe. The local features of the geometry of the universe are primarily described by its curvature, whereas the topology of the universe describes general global properties of its shape as of a continuous object. The shape of the universe is related to general relativity, which describes how spacetime is curved and bent by mass and energy.
(Excerpt end)

Space-time on the scale of the universe is NOT a 'continuous object' with a shape.
Space-time is defined to be the special observer's geometry affected by their proximity to 'mass and energy' but there is no geometry of the universe.
The observer's space-time geometry is background independent with no link to the physical space. It cannot be a real thing.

3.6 Space-time as a Continuous Object

The Space-time as the fabric of space cannot be verified to be real.

There is a simple rule for a thing to be real. It must be measurable by everyone from anywhere.

If there is a privileged observer who claims they have a thing which only they can measure, everyone else should ignore the claim, even laugh at it.

Physics relies on evidence which is obtained by a measurement. Anything which cannot be measured is also a thing which cannot be claimed to be real. Forces can be measured even if not visible.

Real objects do not rely on a specific observer. Others must be able to observe and measure it.

If one person gives a box to another person and they measure it with the same values (every measurement method has a margin of error), obviously they can agree this box is real. (A weight anomaly from the observer's altitude is not relevant because the thing has an attribute measurable by all.)
The crucial requirement for determining the reality of an object is whether it can be measured independently of a specific observer.

There is a multitude of mechanisms to measure point-to-point distances, with the result independent of the person using that mechanism.

Here on Earth, we have 2 position measurements available which are independent of a particular observer.

1) The GPS coordinates have the latitude and longitude planes referenced to the fixed point at the center of the Earth.
The elevation measurement is also referenced to the center of the Earth.

The latitude, longitude, elevation measurements can be made for any point on or around the globe, independent of one observer.

2) The celestial coordinate system has the declination and right ascension (RA) planes referenced to the fixed point at the center of the Earth.

Each observer does a coordinate system transformation for their current position on Earth's surface relative to that fixed point at the center of the Earth.
The distance measurement is also referenced to the center of the Earth.

The declination, RA, distance measurements can be made from any point around the globe, independent of one observer.

Using this Earth-based coordinate system, we can agree on what is measurable and real in our observable universe.
If we cannot measure something independent of a particular observer then it cannot be verified as real.

Right now, the only way to observe or measure anything anywhere in the universe requires an observer on Earth. A space probe leaving Earth can still use the celestial coordinates by accounting for its position relative to Earth.

Probes in interstellar or intergalactic space making celestial position measurements will be challenged to use Earth as a reference for those measurements. Alternately a somewhat fixed point could be used.

A galactic coordinate system has been defined but the moving Sun is its center so it has a restricted basis to be used for references beyond our galaxy.

Observation:

The New Horizons probe traveling to Pluto took an image of Messier 7 for its first calibration of its LORRI camera. That observation required a known location, compared to Earth, for the probe to correctly observe that open cluster in distant space when far from Earth.

We cannot verify any claims of anything as real beyond our observable universe. This statement is simple logic.

The fabric of space, or an instance of space-time, is claimed to be created with the big bang. For a legitimate claim as being real, this fabric must be measurable independent of one privileged observer.

To meet this requirement, a universal coordinate system is required, or one not using Earth as its reference point. Its dimensional planes and a distance from that point must be referenced to a specific fixed point in the universe.

Currently, it is impossible to determine limits on the size of the universe so it must be treated as having no limits (or infinite).
It is impossible to identify a fixed point within any space having no defined limits.

There is no fixed point in the universe to serve as this required reference for the defined dimensions.

If we ever, somehow, define the limits on a finite universe, then it becomes possible to identify a fixed point within that finite volume.

Until that time, it is impossible to define an observer independent coordinate system for the universe.

All measurements for cosmology require an observer to define the dimensions and their reference point in the observable universe.

Space-time is the moving observer's reference frame. When confined to that context, the observer's current position in space is the reference for the space-time dimensions. Space-time is explicitly an observer dependent coordinate system.

Relativity is a background independent theory meaning it has no reference to any physical coordinates in the background of the observer. Though relativity is the basis for popular cosmology, relativity has no role in verifying aposition independent of a special observer.

Therefore the fabric of space cannot be verified as real because it requires a particular observer who must use the Earth for measurements. This requirement for verifying a real thing is crucial.

Cosmology claims to explain the universe using relativity and this can appear possible now within only the observable universe simply because we are all on the Earth which is the basis for our measurements.

The claim of a fabric of space is reaching too far to justify. It cannot be measured beyond a reference to Earth.

The big bang is claimed to create a continuous object, or the fabric of space, which cannot be verified as real.

All cosmological claims based on a fabric of space, as a continuous object, have no real basis when that fabric cannot be verified, by a measurement, as real.

If the fabric is claimed to be infinite in size and it is also expanding, but with means to measure anything about it, then the claim is ridiculous.

3.7 Space-time in the Physical Universe

Space-time cannot be anchored to the physical universe other than through the moving observer whose background independent coordinate system is affected only by local relativistic effects. More about this is below.

The universe is infinite and everything in it can be moving.

It is absolutely impossible to identify a single fixed point in the universe to anchor a proposed coordinate system of the universe.

Any attempt at such a coordinate system must begin with the observer. That means as the observer moves, this coordinate system of the universe is moving as well, so its respective axis planes could be rotating and their references shifting.

Cosmological measurements are based on the observer.

In the big bang cosmology, space-time of the universe was created by the big bang event.

Actually space-time is limited by relativity to the moving observer so the big bang cannot create space-time because only a moving observer has their own space-time, distinct from all other observers
Despite that inherent restriction in relativity, this big bang theory proposes an instance of a moving observer's space-time was created and this 'thing' is a continuous object whose shape can be described by space-time.

If this is a real thing then it must have a physical location in the universe where we can measure it, as evidence.

Even stranger, cosmologists propose a thing called time in this universal space-time was created with the big bang. Time is not a real thing; it is only an incrementing count, typically used for defining a start and end time for a measurement of the delay between them. Isaac Newton considered time as separate from space and he was correct.

The confusion about universal space-time worsens.

The space-time curvature resulting from an observer at a mass is sometimes claimed to be observed at great distances from Earth. Examples are black holes and light bending due to curved space-time caused by a distant large mass like a galaxy.
In each case, the observer must be both where they are on Earth and simultaneously adjacent to that distant mass to get the correct curvature defined by relativity for a moving observer at that adjacent location in the universe. This combination is clearly impossible.

Cosmologists also propose the universe space-time is expanding. It is impossible to identify the context for this space-time within the real universe which must include a fixed point reference for its dimensions. This expansion is also not a real thing.

Expansion involves claiming the space-time reference frame is changing so the observer is measuring positions which can change due to the variable scaling in the observer's reference frame.

Excerpt:

The expansion of the universe is the increase of the distance between two distant parts of the universe with time. It is an intrinsic expansion whereby the scale of space itself changes. The universe does not expand "into" anything and does not require space to exist "outside" it. Technically, neither space nor objects in space move. Instead it is the metric governing the size and geometry of space-time itself that changes in scale.

To an observer it appears that space is expanding and all but the nearest galaxies are receding into the distance.

(Excerpt end)

Observation:
This expansion is the appearance to the moving observer, not real.
In this expansion theory, space-time has its dimensional scaling increasing.

With space-time being background independent this proposed expansion is difficult to grasp when the scope of space-time is extended from only an observer to cover the entire universe.

However, there are known problems with this expansion among physicists.

Excerpt from Wikipedia:

A much slower and gradual expansion of space continued until at around 9.8 billion years after the Big Bang it began to gradually expand more quickly, and is still doing so.

Metric expansion is a key feature of Big Bang cosmology, is modeled mathematically with the Friedmann-Lemaître-Robertson-Walker metric and is a generic property of the universe we inhabit. However, the model is valid only on large scales, because gravitational attraction binds matter together strongly enough that metric expansion cannot be observed at this time, on a smaller scale. As such, the only galaxies receding from one another as a result of metric expansion are those separated by cosmologically relevant scales larger than the length scales associated with the gravitational collapse that are possible in the age of the universe given the matter density and average expansion rate.

Physicists have postulated the existence of dark energy, appearing as a cosmological constant in the simplest gravitational models, as a way to explain the acceleration. According to the simplest extrapolation of the currently-favored cosmological model, the Lambda-CDM model, this acceleration becomes more dominant into the future. In June 2016, NASA and ESA scientists reported that the universe was found to be expanding faster than thought earlier.

While special relativity prohibits objects from moving faster than light with respect to a local reference frame where space-time can be treated as flat and unchanging, it does not apply to situations where spacetime curvature or evolution in time become important. These situations are described by general relativity, which allows the separation between two distant objects to increase faster than the speed of light, although the definition of "separation" is different from that used in an inertial frame. This can be seen when observing distant galaxies more than the Hubble radius away from us (approximately 14.7 billion light-years); these galaxies have a recession speed that is faster than the speed of light. Light that is emitted today from galaxies beyond the cosmological event horizon, about 16 billion light-years, will never reach us, although we can still see the light that these galaxies emitted in the past. Because of the high rate of expansion, it is also possible for a distance between two objects to be greater than the value calculated by multiplying the speed of light by the age of the universe. These details are a frequent source of confusion among amateurs and even professional physicists.

Due to the non-intuitive nature of the subject and what has been described by some as "careless" choices of wording, certain descriptions of the metric expansion of space and the misconceptions to which such descriptions can lead are an ongoing subject of discussion within education and communication of scientific concepts.

(Excerpt end)

Observation:

This section 3 Relativity tries to explain some of the "non-intuitive nature of the subject."
The earlier Section 1 Red Shifts explains their mistake to conclude galaxies have such a high recession velocity suggesting they are older than the universe, or "beyond 16 billion light-years."

There is also no evidence for an event affecting the false expansion at 9.8 billion years ago.

3.8 Motion In Space-time

Some theoretical physicists repeat the saying, "Spacetime tells matter how to move; matter tells spacetime how to curve."

This quote is credited to John Archibald Wheeler.

Excerpt from Wikipedia:

"John Archibald Wheeler was an American theoretical physicist largely responsible for reviving interest in general relativity in the United States after World War II."

(Excerpt end)

That task simply means he was presenting an "interesting'" version for the public, not one directly from Einstein, who never said that quote.

A coordinate system should be referenced to the physical space by the observer and only that observer selects their coordinate system.

Theoretical physicists can accept this laughable statement only by misunderstanding physics, relativity and a coordinate system.

One learns in a physics class about the important forces of gravity, electric, and magnetic, and the result of acceleration from a force on a mass. Objects move by the external forces acting on them.

Objects in the universe do not move using specific coordinates defined by an observer somewhere in the universe.

Excerpts from Wikipedia:

Postulates of special relativity
1. First postulate (principle of relativity)

The laws of physics take the same form in all inertial frames of reference.

General relativity generalizes special relativity and refines Newton's law of universal gravitation, providing a unified description of gravity as a geometric property of space and time, or space-time. In particular, the curvature of space-time is directly related to the energy and momentum of whatever matter and radiation are present. The relation is specified by the Einstein field equations, a system of partial differential equations.

(Excerpt end)

Observations:

Relativity is a background independent theory meaning it never uses physical coordinates. All its field equations are confined to the moving observer's reference frame. Relativity by its very design never affects any physical entity, only the special observer's reference frame whose dimensions can be curved in this mathematical exercise.

When cosmology tries to apply general relativity, the context for space-time is still confined to the special observer's reference frame. This reference frame uses no coordinates in physical space.

Nothing in the universe is a special observer with a non-inertial reference frame using commanded motion
Nothing in the universe moves using a coordinate system.

Everything in the universe moves as a result of forces acting on it. A change in kinetic energy requires a transfer of energy,
A coordinate system is not a source of energy.

The bending of light by space-time has been proposed. The propagation of the synchronized electric and magnetic fields is affected only by the medium, as defined by its diffraction index.
Light will never follow a path defined by a coordinate system.

The use of a coordinate system requires an intelligent observer to command a move described by a coordinate system. To suggest a planet or star will perform a commanded motion is simply foolish.

Some theoretical physicists appear detached from reality. That quote is an emphasis of a manipulated coordinate system lacking a connection to physical space. Space-time requires a person, an observer moving through a gravitational field, to use it as their reference frame.

We are here on Earth, not on every celestial object to command their motions.

Classical physics was grounded in physical space with an established time increment for precise measurements while obtaining valid evidence for experiments to test and verify a theory.

We can measure celestial bodies in motion using our selected coordinate system. They move as affected by external forces.
They do not move according to someone's coordinate system.

Mankind has been observing the universe for a very long time without those bodies following a coordinate system.

3.9 Space-time In Graphics

Graphical representations of space-time curvature are an intentional deception.

This unedited image from NASA will help explain this deception.

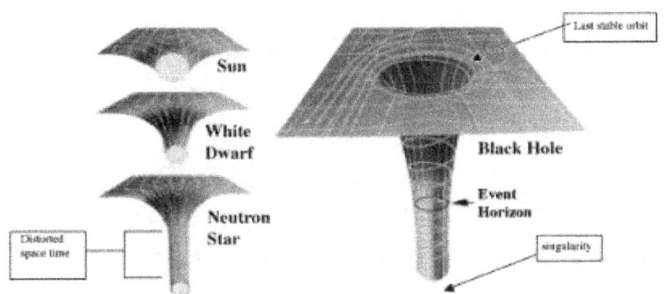

In relativity, when the observer is moving near an object with a gravitational field their 4-dimensional coordinate system will be curved so straight lines in Euclidean geometry are no longer straight. If the user defines their motion using coordinates in the distorted dimensions then their path will not be straight.
This curvature affects the moving observer's coordinate system but no one else is affected.

Einstein's first postulate is "The laws of physics take the same form in all inertial frames of reference."

The left column in the image illustrates how the observer's space-time is curved when the observer is passing by the Sun, a white dwarf, or a neutron star.

For all other observers the Sun, the white dwarf, or the neutron star will be observed using classical physics, such as electromagnetic radiation.

The image is deceptive because there is no distinction between the observer moving past these objects and all other observers.

One could present an edited image to represent the view for all other observers by simply removing those curved graphics for the observer's space-time. At the lower left is the legend "distorted space time" explicitly noting the specific context for this graphic. That edited image removes the deception by showing the real universe, which all observers can observe and measure, and which is not affected by one observer's motion past a particular body in physical space.

The right column in the image has the most blatant deception.

The single arrow pointing to "Singularity" is actually pointing to 2 entities.

1) The physical mass at that location in physical space,

This mass is not shown here though each mass was shown in the left column.

The image could be edited as suggested to remove the graphs from the respective columns; then the mass should be shown here, consistent with the others, to help fix the deception for all observers other than the one moving (i.,e., non-inertial).

2) A point in the observer's reference frame or coordinate system.

The point is not in the image simply because a point has no size.

In basic geometry, the intersection of 2 lines is a point. The point is a specific coordinate in the coordinate system; a simple example of a point in 3-D is X1,Y2, Z3.

In the mathematical exercise of space-time curvature for an extreme mass all the lines of the respective dimensions cross at a point called the singularity.

This singularity is called a black hole though technically it is a black point. There is no hole in anything; it is just a point in a coordinate system.

The deceptive graphic hides this disaster for physics with two simultaneous conflicting entities where one entity is a concept, just a point in a coordinate system, while the other is a physical mass.

For all other observers the mass is present and can be observed and measured and as a mass it is still subject to the force of gravity from other bodies.

Physicists chose to combine these two conflicting entities, resulting in something physically impossible.

The singularity is claimed to retain the mass and its gravitational field. However, this point has no size so the result is a gravitational field coming from a mass having infinite density.

There should be another arrow in the image next to that of Singularity and pointing to the same point but with the legend "Impossible"

There is no such thing as a black hole. This will be explained later.

Probably, if graphical representations of space-time curvature were not deceptive then impossible entities like black holes would go away.

Also, the mistaken claim of remote gravitational lensing should also go away having no justification for a remote curvature.

To present the reality of a proposed black hole, the image for most observers (except for the special observer) who have no distorted space-time, the bottom right should have this note inserted using the Sun's graphic icon (instead of O):

Note:
Milky Way SMBH has O x 4.1 million visible to all other observers.

(end of note)

That simple change to the figure clearly unveils the deception because there is NO huge real mass of that size, , observable by any observer other than the special observer, at that location claimed for that super massive black hole.

This book is about how we observe the universe. The black hole is something we are told is there but we cannot observe it. That restriction makes the concept immediately suspicious.

Tests To Confirm Space-time

The most famous test was the 1919 solar eclipse.
Sir Eddington observed a star during the eclipse, directly at the solar limb for the maximum diffraction through the bottom denser layer in the solar atmosphere.

Other observers noted other stars did not bend correctly. Einstein's prediction expects a deflection proportional to the distance from the star, with the maximum at the limb. No stars other than the one on the limb were affected as expected.

This rigged experiment proves a changing medium of plasma, like in a star's atmosphere, can bend light.
Dr. Dowdye's experiments confirmed light bent as expected at different separations.
It did not confirm gravity bends light.

If it did, there should be distant stars appearing around every large star bending the light of the stars behind it.

For those wishing a thorough explanation, there is a roughly 10 minute YouTube video explaining how light propagating through plasma will bend at a predictable and confirmed angle.

The video is titled "Can Stars BEND LIGHT? General Relativity and Gravity with Dr. Edward Dowdye!"

3.10 Relativity has not been confirmed

For those wishing a thorough explanation, there is a web article roughly describing the lack of observational data confirming relativity. The problem is not disproving relativity. It must have its proof in evidence to be considered valid. The article is titled "common misconception 9 — who disproved Einstein?"

This book contains several demonstrations of the validity of Newton's force of gravity.

A contemporary test of relativity could have been performed using space-time curvature to predict the path of a space probe on a slingshot trajectory. Instead, NASA used Newton's simple gravityequation and the complex path taken using precise timing was just as predicted. Whether NASA attempted a calculation using curvature is unknown, so whether relativity actually failed is unknown. Gravity was confirmed.
There remains no observation where relativity explained a motion when Newton's force did not.

3.11 Better coordinate system for our universe

We are told frequently we live in a 4-dimensional universe.

Unfortunately, the universe has no inherent dimensions and those 4 being offered are wrong for our recording our observations.

Here is a 4-dimensional system for observing our universe, and is better than space-time.

Some say space-time defines or describes our universe.

Some say space-time is our universe and was created with the big bang.
Space-time is a 4-dimensional coordinate system with 3 linear coordinates and 1 time coordinate.
It is easy to say this particular set of 4 dimensions have never been used for a celestial observation in the entire history of mankind.
The space-time dimensions are not practical when observing our universe.

A better approach was initially described over 2000 years ago, so it has been used a long time.

Excerpt from Wikipedia:

Hipparchus (c. 190 – c. 120 bc) was a Greek astronomer, geographer, and mathematician. He is considered the founder of trigonometry.

Hipparchus is credited with the invention or improvement of several astronomical instruments, which were used for a long time for naked-eye observations. According to Synesius of Ptolemais (4th century) he made the first astrolabion: this may have been an armillary sphere which Ptolemy however says he constructed; or the predecessor of the planar instrument called astrolabe. With an astrolabe Hipparchus was the first to be able to measure the geographical latitude and time by observing fixed stars.

Hipparchus also observed solar equinoxes, which may be done with an equatorial ring: its shadow falls on itself when the Sun is on the equator (i.e., in one of the equinoctial points on the ecliptic), but the shadow falls above or below the opposite side of the ring when the Sun is south or north of the equator. Ptolemy quotes a description by Hipparchus of an equatorial ring in Alexandria; he [also] describes two such instruments present in Alexandria in his own time.

Hipparchos laid the foundation for the celestial coordinate system in use today, with its basis in his equatorial ring. With the daily rotation of the Earth the universe is observed as a sphere by everyone so spherical coordinates are correct, not linear coordinates.

(Excerpt end)

All astronomers use the 2-dimensional celestial coordinate system, having 1) an angular value for right ascension (RA) based on an equatorial ring, and 2) an angular value for declination (Dec) which is the angular distance from the equator.

The better 4-dimensional system to replace space-time:

1) the object's measured RA position,
2) the object's measured Declination position,
3) the distance to this object, obtained by other observations (such as parallax),
4) the time of this position measurement.

With this 4-D coordinate system, every object in the observable universe can have its current position described at the time of the observation. Any observer using Earth as the reference can duplicate this position measurement. Time is included in the 4 because a time measurement is required to measure motion of an object by changes in its measured position over time.

Space-time is wrong for astronomy. It was defined for a very different purpose, as the reference frame of a moving observer in the theory of relativity.

Space-time is the moving observer's frame of reference. The values in the 4 dimensions are used in the relativity equations as change-x, change-y, change-z, change-t.

Space-time is for only the moving or non-inertial observer and their dimensions can be curved by a gravitational field. Space-time does not apply to all the others observing the universe. Observations begin with the current measured positions. Changes come from calculations after multiple observations.

When astronomers consistently use the same 4-dimensional system while observing the universe, mistakes are avoided.

Space-time is just wrong for the universe.
Space-time is not real. Any n-dimensional coordinate system can never be real. It is only a framework used by an observer to measure physical positions,
Believing it to be real enables fictional concepts to arise, including black holes and the expanding fabric of space-time.

Replacing space-time with a better 4-D system will lead to a better cosmology, when it ignores an irrelevant special reference frame.

We should also stop getting fictional stories about a multi-dimensional universe.
The universe has no dimensions. The observer defines the coordinate system and its dimensions for their observations.

The simple 3-D Euclidean geometry is always a convenient system for local observations and measurements, like in a laboratory.
Space-time is not appropriate when observing the universe.

Observation:
The widely accepted 4-dimensional system for measuring positions around the Earth is the GPS system.

1) the object's measured longitude position,
2) the object's measured latitude position,
3) the elevation of the object,
4) the time of this position measurement; this item is not required, unless calculating a velocity.

3.12 Conclusion of Relativity

Relativity has no evidence for its claim that gravity moves at the velocity of light. There is conflicting evidence that proves relativity is wrong because protons are measured at a velocity at multiples of c. In all cases, space-time has no solid evidence when it is a valid replacement for Newton's force of gravity. The experimental evidence (by predict and confirm) consistently confirms the force of gravity. Light bends by plasma in the solar atmosphere, not by gravity, not by space-time.

Space-time is not the correct 4-dimensional coordinate system for recording our observations of the universe.
We have been using the celestial coordinate system for a very long time.

Cosmology must have repeatable evidence with no conflicting evidence.

Relativity lacks convincing evidence to justify its claim it is better than Newton's force of gravity.
It claims that both the velocity of mass and the force of gravity are limited to the velocity of light are both falsified.

There is no justification for the continued use of space-time in cosmology.

4 Cosmology Concerns

Cosmology attempts to explain the universe. Cosmology has other concerns beyond those described in the previous 3 main sections.

There is a known crisis in cosmology caused by the uncertainty of a precise value for Hubble's Constant.

This section will describe several other concerns.

4.1 Black Holes

A very large mass results in a very large curvature in the moving observer's space-time. With an extreme mass the curvature collapses to a single point in the space-time coordinate system.
The mass for this black hole's curvature is claimed to be so extreme that light cannot escape. The point is called a singularity to imply it exists in real space not just in space-time. The moving observer has their reference frame collapsed to a point. If the moving observer wishes to move in their reference frame by picking a coordinate, all coordinates are at that point in this curved coordinate system.

To all other observers the mass remains at that location in space.
This was described above in the section 3.10 titled: Space-time In Graphics.

The combination of this mass leaves the visible universe for all other observers; now resides in a geometric point in the special observer's space-time so no one else can see the mass; and yet the point exerts the gravitational field of the mass within the point, at infinite density, is incredible. This is not physics.

This geometric point is given the misleading label of black hole. There is no hole because the geometric point has no size.

The most common use for a black hole is an X-ray source with no visible object.
Nearly every galaxy is assigned one but nearly all galactic cores are congested with dust, gas, and numerous stars so the source is usually obscured.
In this case, a black hole with an extremely hot accretion disk is proposed because with a gravity only cosmology plasma phenomenon are ignored,
There is one verified mechanism for generating X-rays: a synchrotron.

Excerpt from the European Synchrotron Radiation Facility site:

Synchrotron radiation was seen for the first time at General Electric in the United States in 1947 in a different type of particle accelerator (synchrotron). It was first considered a nuisance because it caused the particles to lose energy, but it was then recognised in the 1960s as light with exceptional properties that overcame the shortcomings of X-ray tubes.

In the mid- to late 1970s, scientists began to discuss ideas for using synchrotrons to produce extremely bright X-rays.

The entire world of synchrotron science depends on one physical phenomenon: When a moving electron changes direction, it emits energy. When the electron is moving fast enough, the emitted energy is at X-ray wavelength.

(Excerpt end)

This simply defined mechanism for X-rays has been known for roughly 50 years.

Modern cosmology ignores this known physics and instead proposes a new mechanism never duplicated.

The mechanism is a black hole (an unverified theory) can cause a surrounding disk of material to heat to such an extreme temperature that its thermal radiation extends to X-ray wavelengths.
This mechanism has never been duplicated.

Excerpt from a post at the University Of Cambridge Institute Of Astronomy, about thermal emission:

To be hot enough for the peak of emission to be in the X-ray range the material would have a temperature of around 300,000-300,000,000K.

(Excerpt end)

This proposal is absolutely unbelievable for material in an accretion disk (not fully compressed but loose enough so only friction causes this heat) to reach this extreme temperature and remain intact.

In April 2020, an infamous image was taken of the black hole at the center of the M87 galaxy.

However that donut-shaped object was a plasmoid, not a black hole. This torus of plasma generates synchrotron radiation extending to X-ray wavelengths.
Plasmoids were first observed in a laboratory by Winston Bostick in the 1950's when he coined its name.

A clear explanation of the M87 plasmoid is in a YouTube video titled "Wal Thornhill: Black Hole or Plasmoid? | Space News"

There is no evidence for a black hole. LIGO certainly never detected one.

Nearly every galaxy has an AGN bright in X-ray.
When the known plasmoid cannot be used in the gravity-only cosmology, a black hole is claimed to be there.

A cosmologist must assign a mass to this black hole. In practice, the mass is usually more than the number of stars assumed to be in that galaxy.

This is probably due to the barycenter expectation of a galaxy. The mass at the center must balance all the mass around it.
No black hole in the universe had its mass actually measured. We are just told what is, in some number of solar masses, with no evidence for the claim.

One possible method is finding a black hole in a binary with a star. If the 2 move in an elliptical orbit around the barycenter then a mass could be calculated parameters based on the orbital parameters.

Astronomers keep looking for a star in orbit around the Milky Way black hole. This is nearly impossible. The simple rule with an ellipse is the period increases with the orbital radius.

Astronomers claimed to find a star with a period of 17 years but that period is in the range of our planets. That relatively small number of AU needed for that period cannot be resolved. As the radius approaches 1000 AU then the period is hundreds of years or more.

So far, each combination of a measured radius and period is not a valid ellipse.

Every black hole is assigned a value of a number of solar masses but with no evidence for the claim.

No black hole in the universe has had its claimed mass actually measured to verify the claim.

4.2 Cosmological Model and Dark Matter

A "Finding" is the result of an investigation.

In this topic, a finding is something found or observed during the search for dark matter.

Dark matter is undefined other than by an observation of an unexpected behavior.
Many scientists are looking for dark matter and upon its discovery cosmologists must follow a process of verification because the search is for an undefined entity. It is possible a finding is not the correct finding. The search must continue until a particular finding is verified to be the correct finding, thereby ending the search.
The verification process determines when the search for another possible solution is done.

The verification process entails including the dark matter finding, whatever it might be, in the original scenario and confirming the original observed deviation has been removed by including this finding, claimed to be the cause of the deviation when this finding was not included in the predicted behavior.

a) Definition of dark matter

Excerpt from Wikipedia:

Dark matter is a form of matter thought to account for approximately 85% of the matter in the universe and about a quarter of its total energy density. Its presence is implied in a variety of astrophysical observations, including gravitational effects that cannot be explained by accepted theories of gravity unless more matter is present than can be seen. For this reason, most experts think that dark matter is abundant in the universe and that it has had a strong influence on its structure and evolution. Dark matter is called dark because it does not appear to interact with observable electromagnetic radiation, such as light, and so it is undetectable by existing astronomical instruments.

(Excerpt end)

Observation:

Dark matter is undefined but is found only where a behavior cannot be explained "unless more matter is present than can be seen."
Given that simple criteria for a required explanation, there are 2 simple alternatives:
1) More unseen matter is needed. Or,
2) An unseen force other than gravity is involved.

Cosmologists have simply neglected pursuing (2).

The solution in (2) will be pursued below.

c) Observed need for dark matter

There is an anomaly in a spiral galaxy rotation.

Continue excerpt from Wikipedia:

Early mapping of Andromeda with the 300 foot telescope at Green Bank and the 250 foot dish at Jodrell Bank already showed the H-I rotation curve did not trace the expected Keplerian decline. As more sensitive receivers became available, Morton Roberts and Robert Whitehurst were able to trace the rotational velocity of Andromeda to 30 kpc, much beyond the optical measurements.

The primary claim for dark matter is it explains the unexpected velocities observed in a spiral galaxy rotation.

(Excerpt end)

Observation:

The 'expected keplerian decline' is a mistake because the expectation is stellar motion is like planets in our solar system.

Excerpt of interview with Vera Rubin in a 2006 story titled "Seeing dark matter in the Andromeda galaxy" in Physics Today:

Our 1970 paper included optical observations out to 120 arcmin but did not include the superposed image of M31, or the 1975 radio observations shown in the figure. This composite of the galaxy and velocities emphasizes the extent of the optical image and the "flatness" of the velocities. We found it puzzling that stars far from the center traveled as fast as those much closer to the center. However, we chose not to extend the curve beyond the final measurement by using a decreasing Newtonian inverse square velocity, the common practice at that time. Instead, we wrote "extrapolation beyond that point is clearly a matter of taste."

Isaac Newton showed that the force on a mass at radius r from the center of a symmetrical mass distribution is proportional to the mass interior to that r. High-school students learn that in a gravitationally bound system like our solar system, a planet moves in a closed orbit, such that $MG = V^2 * r$ where M is the mass of the Sun, G is the gravitational constant, and V and r are the velocity of a planet and its distance from the Sun. In M31, the same relation between mass, velocity, and distance holds. A flat rotation curve (V = constant) implies that mass increases linearly with distance from the center. Enormous amounts of nonluminous matter extend far beyond the optical image of M31.

(Excerpt end)

Observation:

High-school students learn of planet orbits but perhaps they should learn the barycenter (the center of gravity) is critical, to avoid the serious mistake of assuming simple 'keplerian' orbits around the Sun is correct for a galaxy.

Our solar system has less than a dozen bodies involved with this barycenter.

A galaxy like M31 has billions of stars in its disk.

This is a mistake to assume billions of stars distributed within distinct arms in the disk move about a galactic barycenter just like the 8 planets in simple ellipses in a limited system of the Sun and 8 planets.

d) Search for Dark Matter

Excerpt from CERN:

Many theories say the dark matter particles would be light enough to be produced at the LHC. If they were created at the LHC, they would escape through the detectors unnoticed. However, they would carry away energy and momentum, so physicists could infer their existence from the amount of energy and momentum "missing" after a collision. Dark matter candidates arise frequently in theories that suggest physics beyond the Standard Model, such as supersymmetry and extra dimensions. One theory suggests the existence of a "Hidden Valley", a parallel world made of dark matter having very little in common with matter we know. If one of these theories proved to be true, it could help scientists gain a better understanding of the composition of our universe and, in particular, how galaxies hold together.
(Excerpt end)

Observation:

References to "beyond the Standard Model" or "extra dimensions" or "parallel world" demonstrate this search is just conjecture (or science fiction fantasy). The definition must be something based in classical physics where evidence by experiment is required.

e) Alternative to dark matter found in 2010

Cosmologists had a choice in 2010 when scientists observed the M31 rotation curve could be explained by the galactic magnetic field meaning the stars were not moving like planets driven only by gravity around the solar system barycenter.

From "Magnetic Fields and the Outer Rotation Curve of 31" the 2010 paper from Astrophysical Journal Letters.

Excerpt:

Observations of the rotation curve of M31 show a rise of the outer part that can not be understood in terms of standard dark matter models or perturbations of the galactic disc by M31's satellites. Here, we propose an explanation of this dynamical feature based on the influence of the magnetic field within the thin disc. We have considered standard mass models for the luminous mass distribution, a NFW model to describe the dark halo, and we have added up the contribution to the rotation curve of a magnetic field in the disc, which is described by an axisymmetric pattern. Our conclusion is that a significant improvement of the fit in the outer part is obtained when magnetic effects are considered. The best-fit solution requires amplitude of [about] 4 microG with a weak radial dependence between 10 and 38 kpc.

(Excerpt end)

Observation:

The rotation curve cannot be explained using dark matter. The best fit is obtained using the galactic magnetic field.

Upon the M31 study's finding, cosmologists could abandon the barycenter assumption and replace it with the magnetic field.

A YouTube video of the paper's presentation can be found by the paper's title: "Magnetic Fields and the Outer Rotation Curve of M31"

Upon the M31 study's conclusion, cosmologists could abandon the barycenter assumption and replace it with the galactic magnetic field.

f) Similar alternative to dark matter explained in 2018

A paper in 2018 concluded the galactic magnetic field drives the galactic rotation and no undetectable dark matter is required.

The paper is titled "Birkeland Currents and Dark Matter" and can be found with a web search.

Excerpt from its conclusion:

An observation that is "anomalous" is one that is inconsistent with accepted hypotheses. In real science this requires the replacement of the falsified hypothesis, not an eighty-five year hunt for invisible entities that will preserve it. The work being presented here demonstrates that the root cause of the now vast collection of observed "anomalous" galactic stellar rotation profiles is the electrical nature of the Birkeland Currents on which those galaxies have been or are being formed.

g) Alternative to dark matter found in 2015

An important conclusion after a study of IC342, a large obscured, nearby spiral galaxy:

Excerpt from the story titled:" Magnetic fields in spiral galaxy arms"

"Spiral arms can hardly be formed by gravitational forces alone," continues Rainer Beck. "This new IC 342 image indicates that magnetic fields also play an important role in forming spiral arms."

(Excerpt end)

Wherever there is a claimed need for dark matter, there is a magnetic field being ignored.

h) Finding dark matter and its verification

There is an ongoing search by many to find dark matter.

We could expect either:

1) CERN actually detects an event with something "missing" as an important finding though as of January 1, 2020 nothing has been found, or

2) Someone else reports a finding of a dark matter candidate.

After CERN or anyone else does an experiment with a result assumed to be this undefined dark matter then the finding must be verified to be the correct solution.

Currently, that verification test is undefined.

The first test should be confirming this finding solves the notable problem of a spiral galaxy rotation curve which is not as expected by the current model.

Dark matter is the accepted explanation for this deviation in rotation, even though dark matter has no description in terms of physics to be measurable and testable.

A finding becomes a candidate for the dark matter solution.

However the candidate must be tested and verified before becoming the accepted solution.

i) Verification

When the spiral galaxy model integrates this dark matter finding then the model's predicted rotation curve should now match the observed curve.

This finding and its integration in the model must be verified.

This integration will be difficult because the spiral galaxy model assumes the stars are in orbits like planets in our solar system. All the planets have at their ellipse's focus the barycenter of the solar system. Even the Sun wobbles around this barycenter. Each body has an individual orbit described by Kepler for the solar system of planets. Dark matter is assumed to change the matter/gravity distribution for the barycenter model to explain the motion of the stars in a spiral galaxy because the stars move wrong, unlike planets.

The description for dark matter is often paraphrased as "we added up all the visible mass and that amount does not explain the motion so there is missing matter and we call it dark matter." This description is explicitly about the sum of all visible mass to determine the center of gravity of that distribution, or finding the location of the galactic barycenter.

Dark matter is being placed in the galactic halo so this model assumes the barycenter is in the galaxy, not in the disk.

After the study of M31 in 2010 (described above) cosmologists could drop dark matter but did not. Dark matter remained an unsolved mystery with many seeking the solution to that mystery.

Cosmologists chose instead to disregard that pivotal conclusion and stay with dark matter and the barycenter model.

Updating this model for dark matter to verify the finding as the correct solution requires a substantial effort.

The current barycenter model must get the dark matter finding integrated with the 1 trillion stars assumed to be in M31. The galaxy model must make its predictions based on each individual star's rotation in M31 about the barycenter, or this simultaneous center of gravity of everything in the spiral galaxy disk (including dark matter). The motion of all stars could be affected also by globular clusters outside the disk or even satellite galaxies. This revised galaxy model must be tested to verify, after dark matter, it correctly predicts the observed rotation curve.

A simple description of the task:

1) Plot the precisely measured orbits of the trillion stars. These are the 'visible matter' in the disk.
For an approximation, gas clouds and dust clouds could be ignored.

This detail is nearly impossible given the limitation of our imaging technology over the distance and some stars are obscured. Also, the time required to determine each orbit with the required precision is also impossible. The Sun's orbit is estimated at 225 million years. That time is required to verify the orbit parameters after the completion of one orbit. Approximations (by not waiting for one orbit) will be required so claims of a successful change to the model can be debated.

2) Determine the distribution in the disk, the location and amount of dark matter required for the updated barycenter of the trillion stars,

3) Determine the distribution in each individual orbit, the location and amount of dark matter required for the updated barycenter of the trillion stars,

4) Verify the orbits of the trillion stars will now follow this changing barycenter location, the instantaneous center of gravity in the disk,

Note the time required to monitor one orbit is millions of years.
We don't know the orbital periods for any stars in M31.

5) If the measured motions do not match those predicted, then repeat steps 2-3-4 again.
6) Repeat as many times as necessary until the correct distribution of dark matter has been identified.

For a trillion stars, much iteration is required to verify the first galaxy.

The critical problem involves the approximations of the individual orbits.
The distribution of dark matter must be approximations as well.
The result is dark matter cannot be verified properly in a spiral galaxy.

For the next spiral galaxy, all these steps must be repeated because dark matter is a proposed physical entity having a distribution which must be defined for each galaxy.

There are two alternatives:

1) they spend very much effort trying to make the barycenter model work and verify the galaxy model using a barycenter is correct for that context, or

2) they admit the spiral galaxy rotates primarily by the galactic magnetic field and discard the barycenter model for a spiral galaxy because clearly the barycenter model applies only to planetary systems.

It will be difficult or impossible to achieve (1) for some number of spiral galaxies.
The failure to successfully complete (1) must result in the selection of (2).

Otherwise dark matter remains an unsolved mystery despite any number of findings.

The selection of (2) means cosmologists must admit there is no dark matter in spiral galaxies.

Eventually, even after the successful observation of dark matter (like being attempted by CERN) the barycenter model for a spiral galaxy rotation must be discarded. It is impossible to verify a finding with that model.

Cosmologists should know scientists in 2010 already identified the removal of dark matter from M31 by applying the galactic magnetic field.

j) Conclusion of verifying dark matter finding

A spiral galaxy rotates by the force of the galactic magnetic field.

The spiral galaxy has no dark matter.

Three separate studies reached that conclusion of no dark matter because a magnetic field was present causing the observed behavior.

Dark matter is the excuse for ignoring a magnetic field.

The barycenter model for a spiral galaxy rotation, based on only gravity and observed matter, must be discarded.

The claim of dark matter being 85% of the universe must be discarded as well.

Until the magnetic field is recognized as driving a spiral galaxy rotation, dark matter will remain impossible to detect and explain.
Valid physics requires a defined object or behavior that is measurable and repeatable by experiment.
Currently dark matter is undefined and non-testable. Dark matter is not valid physics.

Once dark matter is found by a test resulting in a possible 'finding' that newly defined entity must be tested and verified.

Given the main reason for its proposed existence, any 'finding' cannot be verified in the spiral galaxy's measurable behavior by using the model which ignored the real cause of rotation.

Dark matter does not exist and its proposal is not valid physics and should not have persisted so long.

4.3 Neutron Stars

Neutron stars are needed wherever there is a high frequency pulsar. In a gravity only cosmology, a plasma phenomenon is ignored.

The neutron star proposal requires a multitude of neutrons to be compressed into a tiny sphere which rotates very fast and, despite having no electrical charge; it impossibly radiates X-rays like a light-house beacon.

Also, neutrons decay in a few minutes when outside a nucleus.

The proposal any number of only neutrons can remain intact, spin very rapidly, radiate X-rays, and not shatter has never been demonstrated. This is fantasy, not physics.

The likely explanation is a pulse like from an electrical capacitor which alternately charges for a time then the abrupt discharge.

This explanation is available in a YouTube video titled: "Neutron Star" Shatters Theory | Space News"

Big Bang Location And Time

Big bang cosmologists can identify the precise time and location of the big bang event.

The time is known to the public but not the location.

If the precise location were revealed then the reputation and careers of those cosmologists are in danger.

The location can be revealed here.

Excerpt from Wikipedia:

Vesto Melvin Slipher was an American astronomer who performed the first measurements of radial velocities for galaxies. He was the first to discover that distant galaxies are redshifted. He was also the first to relate these redshifts to velocity.

(Excerpt end)

In 1936, Edwin Hubble noted, based on galaxy spectra at the time, our Local Group is on an island separate from the Hubble Flow.

His conclusion was based on the inconsistent red and blue shifts found within the Local Group. For example, M31 and M33 are blue shifted by calcium atoms in the line of sight, or in the intergalactic medium.

The more distant galaxies beyond our Local Group had red shifted hydrogen absorption lines proportional to distance. The intergalactic medium (IGM) was affecting the distant galaxy spectrum.

Hubble recognized this inconsistency for galaxies relative to our Local Group. The effect of the IGM has been known since 1936.

There is a study titled:

"A Remarkably Luminous Galaxy At Z = 11.1 Measured With Hubble Space Telescope GRISM Spectroscopy"

A study of the galaxy with the highest redshift (z > 11) concluded its hydrogen red shift is from hydrogen atoms in the IGM.

Despite some knowing of this IGM cause of red shifts, astronomers continue to assign galaxy red shifts to a velocity; this is a mistake with consequences.

Because the red shift is not from the galaxy both its assigned velocity and derived distance cannot be justified. The universe expansion cannot be justified on this basis.

A galaxy spectrum can reveal nothing about a transverse velocity. That motion must be determined from monitoring the motion of individual stars over a span of time long enough to measure a change in its position.

As far as I know, this exercise has been performed using HST for only 3 galaxies: M31 and both Magellanic Clouds. M31 had no observed transverse motion in the selected individual stars. Both Magellanic Clouds have an observed transverse velocity for the selected stars.

Galaxies with a blue shift are within the Local Group or in its vicinity.

One could create a 3-D plot of the velocity vectors for all the distant galaxies.

At this time there are no known galaxies other than the Magellanic Clouds with a transverse velocity. For all others, the galaxy motion is assumed only in the direction of Earth.

All those vectors converge on a single point in the universe, the planet Earth.

The big bang event is assumed to provide the initial velocity for all objects in the universe. Distant galaxies are moving in the direction from their origin.

The precise location for the big bang event must be at the planet Earth. This follows from the mistake with red shifts.

The time of this event is also very precise.

From Wikipedia:

In physical cosmology, the age of the universe is the time elapsed since the Big Bang.

The current measurement of the age of the universe is around 13.8 billion years (as of 2015) – 13.799±0.021 billion years within the Lambda-CDM concordance model. The uncertainty has been narrowed down to 20 million years, based on a number of studies which all gave extremely similar figures for the age.

(Excerpt end)

The narrow precision of this age is laughable.

A meeting of cosmologists in July 2019 could not agree on a precise value of Hubble's constant so it remains uncertain in a "range of a few percent."

An uncertainty of only 20 million compared to 13.8 billion cannot be justified when the rather uncertain Hubble's constant is crucial in this age estimate.

Right now, big bang cosmologists can describe very precisely the time and location of the big bang event as:

Time: 13.799 billion years ± 20 million years

Location: planet Earth

If this conclusion were widely publicized, big bang cosmologists could have a credibility problem.

The Earth centered universe of Ptolemy and others was discarded long ago.

Big bang cosmology returned the Earth to the center of the universe.

If everyone knew this result, perhaps its adherents would fix their mistakes.

4.4 Big Bang Sequence

The Big Bang Theory is Bad science because it is also Bold and Baseless.

Bold - we have the bravado to claim we understand the universe well enough to describe how it was created.
Baseless - there are far too many unknowns for the theory to be even a rough guess
there is no valid basis for most assumptions.

1. It is impossible to know when it happened.

Cosmologists assume an incorrect conclusion that the universe is expanding. Hubble's Law is a mistake treating absorption lines as actual velocities, so there is no way to know the start time. It is always a guess.

2. It is impossible to know the initial conditions.

Cosmologists talk of starting with a singularity but it is impossible to know anything about what existed at the start time.

To even begin developing a scenario we must know what is present at the beginning.

Any mass at the start could have a temperature > 0K. That is important for the amount of energy at the start, present with the initial amount of mass.
There is no basis for determining the amount of mass and energy at the start.

That is required to define how the start must change to become the end.

3. It is impossible to know the final conditions.

Our technology limits the extent of our observable universe.

Approximately 10% of the sky remains difficult to survey as extragalactic objects can be confused with stars in the Milky Way.

As our technology improves we should expect to observe more objects that are too dim right now. The big bang must define how those initial conditions become the final conditions - which are unknown. Therefore the big bang has no valid end goal.

3. How the universe works now is not fully understood.

3a. Cosmologists still cannot explain our universe when dark matter and dark energy are still around as place holders until a real explanation is found.

3b. The theory arose partially to explain the claimed expanding universe.

It is not expanding. These claims of everything moving away from Earth are based on a mistake with red shifts, or a misunderstanding of the Doppler Effect.

This mistake was explained in Section 1 (Red Shifts).

It is impossible for the big bang to create the correct amount of these unknown dark entities, if they truly exist (they don't).

4. It is impossible to know any intermediate conditions.

5. The theory requires an unverified span of time from start to end. We cannot know any intermediate steps.

For example, our universe has a number of intergalactic structures that span billions of light years. It is impossible to know how many years were required for them to form when we don't even know if they formed in place or whether all the pieces moved into their positions along unknown paths.

One current assumption for the big bang sequence involves the transient existence of antimatter.

Anti matter consists of atoms having antiprotons and positrons.

In our current universe there is no antimatter; only infrequent antiparticles might exist very briefly before they are destroyed by normal particles.

Cosmologists propose the big bang sequence creates both matter and antimatter but the matter wins and the antimatter disappears. To claim this intermediate step actually happened with no observation, the CMB is claimed to be evidence of that step having annihilation of antimatter, a transient event with no basis for its proposed time in the sequence.

Attempts to measure the CMB neglected to account for both the Earth's oceans and the strong signal from the Milky Way galaxy in the foreground.

When the galactic foreground is much stronger than the background, extracting the correct background signal is difficult, probably impossible.

Dr. Pierre-Marie Robitaille has published several papers and presented several YouTube videos explaining his analysis of the equipment and analysis of the different attempts to measure the CMB.

One of the first in a series of YouTube videos about the CMB can be found with: "The Big Bang & Microwave Background - The Early Years"

There is no valid evidence for the CMB.

There is no evidence for any intermediate steps in the proposed sequence. All steps are just conjecture.

6. If cosmologists don't know what's in our universe and don't understand how our universe works now, and they cannot predict how everything changed over the course of billions of years to get "here."

The big bang theory is bold and baseless - and is just conjecture, and is also a bad theory, unable to be tested.

There was no big bang.

4.5 Cosmological Model Requires Dark Matter

This is a critique of the Lambda-Cold Dark Matter model by using a document describing its software.

The excerpts are from this academic paper:
"Adaptive Techniques for Clustered N-Body Cosmological Simulations"

Dark matter is the foundation of modern cosmology, not relativity as one might expect, after reading this document.

Neither space-time nor relativity occurs in this document about the model.

Perhaps the Lambda-CDM model is not truly important to how we observe the universe.
However this model represents how well we understand the full extent of the universe we are observing.

Excerpts and comments will follow before a final conclusion.

The document describes the techniques used in the cosmological model.

Excerpt from document:

Constraints on cosmology are tightest on scales of tens of megaparsecs and larger
due to observations of the Cosmic Microwave Background, giving us detailed initial conditions; however our knowledge of the non-linear evolution of the Universe
and of the properties of galaxies is still imperfect, as the detailed properties of
Dark Matter and of Star Formation (SF) remain only partially understood. On the other hand, simulations of large volumes of the Universe, and of individual galaxies at high resolution have been fundamental in putting the standard hierarchical, Cold Dark Matter dominated model, on a robust footing.

(Excerpt end)

Observation:

The CMB defined the initial conditions.
The CMB is noise from the Earth's oceans, as explained by Dr. Robitaille, so this model has invalid initial conditions.

'detailed properties of Dark Matter and of SF remain only partially understood' but in an apparent contradiction, this 'model is on a robust footing.'

These poorly understood facets can be crucial.

The model is hierarchical. Dark matter is part of the foundation for the start of any structures created by the model.

Excerpt:

For example, understanding the development of structures at very high redshift will present different parameter and algorithm choices than simulations that model the observations of current large scale structure.
(Excerpt end)

Mismanagement of red shifts is the major pitfall for modern cosmology.
Not all red shifts are the relative velocity of the object. Only an emission line shift is usually accurate for that atom. This line is used to check the rotation of hydrogen gas clouds in a spiral galaxy.
However a quasar with a relativistic hydrogen atom is definitely not moving at that atom's velocity.

The hydrogen absorption line red shift was found to be roughly proportional to distance by V.M. Slipher before 1920; this shift is due to neutral hydrogen atoms in the intergalactic medium. Intervening hydrogen gas clouds will increase this red shift making the proportion inaccurate.

There is a significant challenge to explain high red shifts in large scale structures.
Quasars near galaxies in a cluster have their anomalous velocity.

Chandra has imaged hydrogen gas clouds in some clusters so these clouds could disturb a more uniform pattern. Many of the blue shifted galaxies in the universe appear to be in the vicinity of the Local Group (including those roughly in the direction of Andromeda M31 due to calcium absorption line blue shift from ions moving toward the Milky Way). The result is nearly every galaxy is receding and the most distant ones are faster (increasing red shift by distance).

Galaxies which appear to be at a similar distance could appear to have the same recession velocity but there is no observed source for the force causing this velocity in this consistent direction (away). Since electromagnetic forces are ignored, dark matter as a weak force of gravity is claimed to be distributed in the area to 'pull' objects around.

Excerpt:

When simulating dark matter and stars, the goal is to understand the evolution of a smooth distribution function that closely approaches a Boltzmann collisionless fluid. As the N-body code is sampling this distribution using particles, a more accurate representation of the underlying mass distribution is obtained if the particles are not treated as point masses, but instead have their potential softened. Softened forces are also of practical use since they limit the magnitude of the inter-particle force. Typically, the softening length is set at the inter-particle separation at the center of DM (Dark Matter) halos.
(Excerpt end)

A halo definition from Wikipedia:

According to modern models of physical cosmology, a dark matter halo is a basic unit of cosmological structure. Thought to consist of dark matter, halos have not been observed directly. Their existence is inferred through observations of their effects on the motions of stars and gas in galaxies. Dark matter halos play a key role in current models of galaxy formation and evolution. Theories that attempt to explain the nature of dark matter halos with varying degrees of success include Cold Dark Matter (CDM), Warm Dark Matter, and massive compact halo objects (MACHOs).

(Excerpt end)

Observation:

The model assumes matter is just a fluid. The only way it can develop into a structure with only gravity within the fluid is with an outside force which comes from DM halos.
An alternative is any ionized atoms or free protons and electrons could behave as plasma, bringing into the fluid electric magnetic fields from plasma in motion. This alternative is ignored.

Excerpt:

In order to efficiently and accurately simulate a portion of an infinite Universe, we perform the calculation assuming periodic boundary conditions. Because of the long range nature of gravity, the sum over the infinite number of periodic replicas converges very slowly.

(Excerpt end)

Observation:
When limited to weak gravity the outcome is reached "very slowly." The age of the universe is assumed over 13 billion years but maybe even that is not enough time for this slow process.

Excerpt:

Despite being a small fraction of the energy density of the Universe, baryons play a significant role in the evolution of structure. Not only are they the means by which we can measure structure (e.g. via star light), they can also directly influence the structure of the dark matter via gravitational coupling. Therefore following the physics of the baryonic gas is essential for accurate modeling of structure formation.

(Excerpt end)
Observation:

"[Mass] plays a significant role in the evolution of the structure of dark matter through coupling."

Clearly when electromagnetic forces are ignored, leaving only gravity, dark matter must have the primary role in the evolution of a structure. Hence its estimate of 85% of matter in the universe even though none of the dark matter can be detected. At only 15%, matter has a minor role.

With 85% invisible, this is certainly a magic trick.

Here is a very simple analogy:
I provide a recipe for a cake. 15% of the ingredients when mixed can be treated as a fluid.

After adding a secret ingredient, amounting to 85% of the final, total ingredients, the batter will thicken to achieve the desired taste and firmness. I can tell you nothing about this secret ingredient though you wish to make a delicious cake.

This secret ingredient, dark matter, has no defined characteristics, not even the composition of its particles; certainly it has no defined quantity to be distributed at some density.
This is an utterly useless recipe.

This is a very ambitious cosmological model, while providing no defined mechanism to explain how the secret stuff is so much more important than what we can observe and measure,

With this sophisticated model, cosmologists claim we understand completely how the universe works. That claim cannot be justified.

Excerpt:

 Hence we need a prescription for where the stars are forming.
Furthermore, it is clear that star formation is a self-regulating process due to the injection of energy from supernova, ionizing radiation and stellar winds into the star forming gas.

(Excerpt end)

Observation:

Again something outside this 'structure' is required including radiation and winds.
Unfortunately the source (stars and supernovae) of those must arise before that source can affect this fluid.

Excerpt:

We have implemented the "blast-wave" and "superbubbles" feedback models. In both models SF occurs in high gas density regions and the time distance scale for energy injection into the gas is then determined by physically motivated models. The "blastwave" prescription follows an analytic model of the Sedov blast wave and it has allowed us to successfully model a number of trends in galaxy populations.

(Excerpt end)

Observation:

Electromagnetic forces with or without plasma can bring structure more effectively with normal physics. The interaction of attractive and repulsive forces can achieve equilibrium. Gravity alone must lead to collapse. Instead the chaos of "blastwave" events and 'winds' are assumed capable of bringing structure.

Excerpt from the conclusion:

With these features, we can bring to bear the computational resources of many 100s of thousands of processor cores on the highly clustered, large dynamic range simulations that are necessary for understanding the formation of galaxies in the context of large
scale structure.

(Excerpt end)

Observation:

They achieved their goal of simulating 'the formation of galaxies in a large structure.'
Whether this model's outcome has any relevance to the real universe is not certain.
The critical fundamental assumption is dark matter determines the formation and evolution of the structure and yet dark matter is 'partially understood.'

This model is obviously a work in progress.
However its reliance on undefined dark matter must impede any progress.

The Lambda-CDM model is a failure at a true representation of behaviors on the scale of the cosmos.

This model is the wrong context for how to interpret our observations of the universe.

4.6 Conclusion of Cosmology Concerns

Cosmology has several concerns (or claims) which can be dismissed by the lack of evidence: dark matter, dark energy, universe expansion, and the big bang.

The cosmological model is problematic having no justification for its purpose – to explain how the big bang created the universe. It can serve no useful purpose now.

5 Cosmology in Chaos

After reading the previous sections in this book one could conclude there is chaos in cosmology.

Claims are being made with no evidence. This is not proper science with accountability.

Perhaps chaos is the wrong word but the science lacks discipline.

Cosmologists have used the word crisis for their inability to agree on a new value for Hubble's Constant. That accepted crisis persists. To suggest chaos is reasonable.

Section 1 concluded with a way to fix Hubble's constant and to resolve the crisis.

There are few examples of a celestial spectrum in this book.

The conclusions drawn were based on a small number of spectrographs along with written descriptions of a spectrum. One spectrum in this book revealed one of the mistakes in the study claiming confirmation of the accelerating universe expansion but it did not given its many mistakes. In 2019 this author noted the flaws in that study and posted the details on January 21, 2020. If others noted the errors as well, the public has not been informed. There is no mechanism to distribute such observations. That mistake and others are explained in this book, but this is not a mechanism to cause mistakes to be addressed by the broad scientific community.

Wikipedia is the conduit for much public information. For a galaxy or quasar, only a relative velocity is published, either as a z value or in km/s. The spectrum for that calculation is never provided. The spectrum was probably published in a research paper within an academic journal, so it would be available to those having access to them.

This research uncovered several, like from an astronomer using a large telescope in France, from a University of Oregon lecture's material, or from Halton Arp's book Seeing Red.

Apparently the scientific community accepts the claimed velocity value with no means of verification.

Arp's book and its spectrum data revealed a quasar has 2 red shift mechanisms. Arp used the lower value and apparently the higher value is used consistently since then. As noted in that section of the book, the quasar red shift, either value, is not the velocity of the quasar.

Without publishing the spectrum details with their claimed velocities, there is no accountability here.

LIGO has made claims of detecting gravitational waves for more than 4 years without ever having one claim verified by an independent observation for confirmation.

On December 16, 2019 the National Science Foundation, who funds LIGO, was notified by e-mail of the November 10 test and confirmation that an earth tide wave is detected by LIGO as a gravitational wave. The subject line in the email: LIGO accountability.

No change could be observed in either LIGO's operations or its reporting of more unverified gravitational waves, each with a probability assigned to the specific binary pair of the claimed detection.

Einstein's theory of relativity is claimed to be correct, with its space-time being a reaction to a gravitational field for only that special, moving observer. However, Isaac Newton's gravity has substantial evidence backing it.

Newton's gravity equation is used for a NASA slingshot trajectory calculation, not Einstein's field equations.

However that evidence required for relativity is lacking. Perhaps the evidence is the 1919 eclipse which was inconclusive because only the single star directly at the solar limb was measured precisely. Later experiments confirmed stars off the limb do not bend as predicted by Einstein. The plasma in the solar atmosphere bends light not gravity.

Relativity is based on a moving observer. Though we are all on the moving Earth we do not observe our universe in the context of relativity's special observer.

The big bang theory remains popular despite no evidence for anything in that complicated theory.

This is not how the scientific method works. Evidence is required for claims in science.

Science is the accumulation of knowledge. At any moment, new evidence could overturn previous conclusions which were verified by the evidence available at that time.

Several fields of science seem to possess inertia. Existing assumptions are repeatedly confirmed without innovation. It takes many conflicting observations to cause a change in course.

Hannes Alfven was awarded the 1970 Nobel Prize in Physics for bringing to light the importance of plasma physics. His work and others, much of it done in a laboratory for independent verification of observations, has been unable to divert astrophysics from its commitment to a gravity only cosmology. That unwillingness to accept change resulted in the mistake of dark matter, which is needed when magnetic fields are ignored, as in a spiral galaxy rotation or in the structure of a plasma filament.

The progress of science requires evidence, to either confirm a new theory or cause a correction in an old theory.

Possible Reaction to This Book

There might be a reader who gets to this point and asks an obvious question:

Why did no one else notice "that" ?

The answer is probably "consensus" or "group think"

In 1922, Edwin Hubble had evidence using a Cepheid that the Andromeda nebula was far away from our Milky Way, becoming the first known external galaxy. If the spectrum for either the galaxy or Cepheid were ever public the are not now.

At the beginning everything must have been on paper.
 That was only about 100 years ago. Around the same time, Slipher concluded a galaxy red shift is its velocity. This conclusion was made before many more galaxies were measured.

In 1933, Zwicky could not explain the red shift velocities in the Coma galaxy cluster so he assigned the cause to dark matter. This cause had no definition beyond "it did it" which is not testable science,

In 1936, Hubble concluded red shifts beyond the Local Group were different.

In 1968, Rubin solved the known problem with the rotation curve of M31 being inconsistent with the orbits of the planets around our Sun.

There was no actual reason why they should match. At the time, the Magellanic Clouds were considered irregular galaxies. Only in recent years were they found to be distorted spiral galaxies.

This analysis involved only one spiral galaxy, M31.

M31 has a fairly dense flat disk of stars and dust in several arms rotating around a large bulge. This complexity does not look like our dispersed set of 8 planets.

She simply assigned the cause of the rotation anomaly to dark matter, which was Zwicky's already accepted explanation for an unexplained motion.

She did not consider the galactic magnetic field.

The date when the M31 magnetic field was first measured could not be found.

These conclusions in the 1930's had very little data for their basis.

Unfortunately, once many agree on a conclusion, regardless of its evidence, that conclusion is difficult to replace. Later observations will look for confirmation of the consensus position. Overwhelming evidence is required to overturn what has become the consensus, or "group think."

A child with a table of moon phases and their dates could observe the frequent match with LIGO GW dates.

That child would be ridiculed like "they would have checked for that!" but there is no evidence LIGO ever noticed this correlation. Its design did not mention the anticipation of the periodic earth tides. The first detection was on the date of a perigee, a cycle which few know of. If no one on the LIGO staff thought of a perigee, then they missed the obvious conflicting evidence for their surprising first detection. After no test of the system it is impossible to know the extent of truth in their first merger description. After that first detection with only one group expressing doubt, they were confident in their system, despite having no real evidence for that first claim and for all that followed.

Science is not self-correcting as some claim. There must be a commitment to follow the evidence, not to follow the consensus. Disagreeing with an authority is a risk in any social group.

This small book cannot provide overwhelming evidence to overturn several popular theories. The glaring problem is some of those theories have little or no evidence justifying them. They persist only because they remain popular careers can depend on stagnation of the basics

This book finds fatal flaws in one of those basics, the red shift.

Science requires evidence, not a vote.

Unfortunately for progress a systemic change is required to ensure new efforts are managed in a new manner.

Astronomical data before 1980 had to be on paper because a personal computer did no exist yet.

Since then, there is a wave of new data. When it became digital is not public.

Recent Gaia data for a billion objects were in digital form.

Wikipedia posts many numbers with no data to justify them.

One must suspect many numbers still come from 1 or more sheets of paper, perhaps scanned images of an archive as just a digital copy.

Probably most of the archived galaxy spectrograms remain only on a sheet of paper in a file cabinet.

The unknown is by the time of HST, there is more digital signal processing than photographs from film. Hubble must have an impressive digital catalog somewhere.

Section 1 of this book observed many cases of no public data.

Eventually astronomers, and researchers like the author, need a public archive of astronomical data.

The funding, site, design, staffing and computing resources could be difficult to obtain.

Until that dream is realized, cosmology will be crippled by the lack of public data.

About the Author:

The author is a retired electrical engineer. The career spanned many roles but consistently, claims of performance always required evidence, and usually documentation. For one example, laser interferometers were used for an independent verification of accuracy and repeatability, with a document including the data.

This book arose from the author's initial double goal: 1) explaining the red shift problem which resulted in the illogical connection between a red shift velocity and a distance in Hubble's Law, and 2) explaining how LIGO reports gravitational wave detections when a GW does not exist.

Relativity became the third main topic because the saying "matter tells space-time how to curve and space-time tells matter how to move" is impossible to reconcile by someone familiar with physics and the 3 inverse-square forces, and with managing layers of configurable coordinate systems.

(End about)

Dark matter arose because magnetic fields were just ignored. Several published studies reached that conclusion, but it persisted.

This ignorance of electromagnetic forces apparently arises from having the mistaken belief gravity alone drives every celestial behavior. Gravity is actually weaker than the electric and magnetic forces, and nearly the entire universe is plasma, or electrically charged. Perhaps relativity distorted the perspective of cosmologists in their pursuit of understanding their observations.

LIGO continues operations despite never having evidence to justify any claim. NSF was notified of a problem with LIGO GW claims, but this bad science continues unimpeded.

Allowing such mistakes to persist is not proper science.

Some careers can tolerate unverified claims. The author's career did not. The author is astounded by this tolerance of lack of evidence in the science of cosmology.

It is very rare to find a public spectrum for a star, galaxy or quasar. This prevents independent research using the original data. There is also no mechanism for accountability for wrong conclusions.

A simple analogy:

An international sporting competition is performed. Only the judges can observe the events. Awards are given to some athletes and to some teams. There are no scores for the public to compare to the individual performances. There is much publicity for the winners.

Such a competition would be meaningless and no one would be willing to participate.

However, cosmology performs its measurements and conclusions using archived spectra in a similar ridiculous context.

Apparently, the spectrum measurement of a celestial object is treated as intellectual property of that observer and is not freely available in the public domain.

This book describes some concepts which should have been dropped by now, simply due to the lack of evidence.

Conflicting evidence should force old theories to reconcile old conclusions with new data.

The intent of this book is describing several fundamental concepts in modern cosmology having problems. Many other concepts are out of scope for this context. The result is a book with only 5 main topics. An archive of the author's deeper research is available on his web site, cosmologyview.

The book's expectation is the reader gets clear descriptions for a new, appropriate basis on how we should be observing our universe.

6 Final Conclusion

The red shift explanation is main purpose for this book.

Changing the units of Hubble's Constant can begin the necessary changes in the correct application of red shifts.

Correcting the mistake with red shifts affects nearly everything in current cosmology from the false big bang to the unsuccessful search for the false dark energy which must drive the assumed expansion of space.

Many theories about the universe resulted from simple mistakes which persisted with no correction.

The false dark matter must be ignored, rendering the current Lambda – Cold Dark Matter cosmological model obsolete.

Cosmology must start over with a new perspective of knowing little with certainty of the motions of everything beyond our Local Group.

We can measure galaxy distances using Cepheids.

For all galaxies without them and for all quasars:

We can place them on our observable universe's sphere but we cannot reliably measure their distances or their relative velocities.

For all galaxies and quasars, we cannot easily measure their velocities.

We can observe distant objects but we don't know for sure whether any are moving.

The current perspective is one of confusion with everything rapidly moving away from us here on Earth.

That perspective puts Earth at the center of an expanding universe which will eventually fade away into the distance.

That depressing perspective is actually wrong.

We need some humility.

We have limitations on what can be measured at great distances.

Advances in technology will gather more data about what we can observe. Space probes enabled our deeper understanding of our solar system. New space probes are observing particular wave length bands.

Our new perspective becomes one of wonder from the distribution of the observed objects spanning an impressive scale, in our universe observed from here on Earth.

In our Milky Way, our Sun will take millions of years for one orbit. This motion is too small to observe in a human lifetime.

Our nearest large galaxy, M31 in Andromeda, has its distance measured at over 2 million light years. It will take many human lifetimes to measure any motion by M31. An object moving at the speed of light moves only 1 light-year per year. The measurement with the precision of 1 light-year is difficult beyond our Local Group.

Deep field images revealed countless galaxies extending to the limit of our telescopes, when they become too dim by their distance.

It is literally impossible to detect motion within this multitude of galaxies on this distance scale.

We can observe intergalactic structures spanning billions of light years. It is impossible to observe the sequence for how they formed at their location, and how long it took to achieve that appearance.

The big bang theory arose from the arrogance we completely understand our universe, and are capable of describing how the current observable universe resulted from a single event at a specific time, following a specific sequence of its evolution. This theory simply ignored our gaps including the accepted dark matter and dark energy (both are conjectures lacking evidence) which reveal how the theory lacks a firm basis. We do not fully understand our universe. We cannot even see all of it due to the zone of avoidance.

We must admit predictions of anything beyond our Local Group are difficult to confirm, given the long time required for the necessary observation.

Progress in science is defined by predictions from theories getting evidence for their verification.

Conflicting evidence refutes the theory, and should prevent its presence being a distraction.

If the invalid theory persists, then progress is misdirected by holding onto wrong ideas.

Theories lacking any evidence are just attempts to improve upon conjectures.

Unfortunately, cosmology stopped demanding evidence so some baseless conjectures remain and any progress is disturbed, or even misdirected.

The universe is a visual wonder. Unfortunately we are limited, by our lifetime, to how much we can learn about many distant objects. Measurements can span generations but such an effort requires a substantial commitment to achieve a conclusion.

Cosmology has an apparent difficulty with managing the observer's context. A Doppler Effect in the line of sight was incorrectly used as a 3-D velocity.

Perhaps someday, cosmologists will be enlightened and return to proper science, which requires valid evidence for claims.

7 References

The references in the book are available as clickable links from a page in the author's web site.

1. Start web browser
2. Go to this site: www.cosmologyview.com
3. Make sure the browser is on the correct home page:

Cosmology Views

4. Scroll to near the middle.
5. Select: **Books by the author**

This page presents information for each book.

Locate the columns for this book.

6. Locate: **Observing Our Universe**
7. Below it, locate the date of this book's edition:

07/30/2020 References

8. Select: **References** after the correct date.

The selected page will list the references in the book by page number, with a link to that reference.

Each link indicates whether it is to a pdf, a YouTube video, or a URL link to a web page. The user is aware of what the browser will do with the link.

www.ingramcontent.com/pod-product-compliance
Lightning Source LLC
Chambersburg PA
CBHW071356210526
45465CB00001B/118